Fort Hickok

Fort Hickok

✦

A Green Community

Waste Not, Want Not

Ivan D. Matthews

iUniverse, Inc.

New York Lincoln Shanghai

Fort Hickok
A Green Community

Copyright © 2007 by Ivan D. Matthews

iUniverse books may be ordered through booksellers or by contacting:

iUniverse
2021 Pine Lake Road, Suite 100
Lincoln, NE 68512
www.iuniverse.com
1-800-Authors (1-800-288-4677)

Because of the dynamic nature of the Internet, any Web addresses or links contained in this book may have changed since publication and may no longer be valid.

The views expressed in this work are solely those of the author and do not necessarily reflect the views of the publisher, and the publisher hereby disclaims any responsibility for them.

ISBN: 978-0-595-45600-0 (pbk)
ISBN: 978-0-595-89901-2 (ebk)

Printed in the United States of America

Contents

Fort Hickok may have never existed as a military post but this conceptual community has been designed to draw tourists to the locale and to demonstrate the practice of living a comfortable yet modest and sustainable lifestyle. After a trip to the Amana Colonies of Iowa in the spring of 2007 I have brought together my interest in old military forts, alternate energy, and affordable living to design this development. As with the Amana Colonies, Fort Hickok is intended to maintain a degree of self sufficiency by employing the "We can do it." principle and common sense. Conserving motor fuels by producing things locally is the underlying theme, and this approach to serving our needs will also help to reduce our carbon foot print. So let's go green in a practical and sustainable way by considering the ideas in this conceptual fort.

People seem to be naturally inquisitive, and for them to see cheese being made, or honey being processed, wheat being ground into flour, or corn into cornmeal is an entertaining and educational experience. Such was the trip to Amana, nothing fancy, no bright lights, but a view of a simple and honest way to live. The tenacity and ingenuity of the pioneers of that community (actually seven small villages) is something to envy. Behind the Fort Hickok concept of local industries, attractions and entertainment opportunities is also the principle of driving less to save fuel, which was very evident in early Amana because the villages were self sufficient in many ways.

The window dressing is intended to attract tourist trade, but there are genuine and practical objectives to be served, the primary one being to conserve motor fuel used in transportation and a secondary one being to conserve the amount of energy needed to operate a home or living space, or a factory or commercial activity. An additional side benefit would be locally grown wholesome food with content that can be verified and that may contain less in the way of preservatives than the typical store bought products.

The possible benefits to people, be they residents, visitors, or even nearby customers of this small community, are sometimes obvious and sometimes subtle and hidden, so I intend to point out the virtues of the approach to living green. This development, although it may involve a relatively small number of participants by modern urban standards, is rather sophisticated and comprehensive, and hopefully is open ended, so the various facets of the designed intent will be categorized and discussed as a related group to the extent that is possible. The overall design is modular enough so that all or only a part of it could be utilized in other villages to achieve the desired effects, and the simplicity of the concept is perhaps the greatest virtue. What you will see is a blend of the old and the new with the intent of having the best from both worlds.

In the end I hope to convey to you the reader an alternative to the customary struggle for existence and to challenge you to conceive your own green community and to follow that dream. If actually creating one of these little villages and helping to reduce the impact on the environment is as much fun as chasing the dreams then it is a sure fire way to have a pleasant journey. For those of you who haven't visited the Amana Colonies and the various ethnic communities in Iowa I urge you to go there and see for yourself what people can do if they work together as a community towards a common goal. The colonies today are a mere shadow of former times but the approach and the philosophy speaks for itself with its simplicity and honesty. Fort Hickok could do well to emulate this wonderful place in Iowa.

Ivan D. Matthews
Abilene, Kansas
May 2007

Overview

There are reasons for including a particular business activity in the project, and I will try to explain the rationale behind each activity in terms of energy conservation, self sufficiency and other virtues if applicable. Some activities are ordinary and some are special, but the thing to look for is the reason why an activity is a part of the overall design and how it contributes to energy conservation and self-sufficiency.

Let us first examine industrial activities that are intended to reduce the fuel needed to bring products to the local consumers and to provide local jobs to those who can ill afford to commute to another town for work. This is particularly important for those who are low on skills and do not have an economical and dependable means of transportation to commute each day to work. The jobs being offered may not pay as well as those in the larger cities but for those who aren't qualified to do better it is still a good deal. Some of the activities are seasonal and some can be done year round, and hopefully a person involved in a seasonal or part time activity can find other work in the off season to round out needed income. Part time jobs are attractive to retirees and to those who want only a second income for the family, or for students who either cannot work full time or do not want to work full time.

Chapter 1 Traditional and New Field Crops

Most of the crops listed below will grow well in Kansas, and the methods and equipment already exist, but the biomass energy crops may require training and new types of equipment to plant and harvest them. As to woodlots grown for fuel, people have been accustomed to harvesting wood for decades so it should not present a major learning or methodology hurdle. However, growing wood with the intent of using it for motor fuel may take some adjustment and change of mindset.

1. Wheat—The primary intent is to have grain for a local flour mill or as a cash crop. Wheat can also be used as an ingredient for various animal and human foods.

2. Oats—Used to make animal feed and human food.

3. Barley—Used for animal and human food products.

4. Corn—The most versatile of the cereal grains, corn can be used for animal or human food products.

5. Milo or sorghum grain—This crop is also used primarily as animal feed but some people eat products made from it. It is a handy crop to grow in areas where there is not sufficient rainfall or irrigation water to grow field corn.

6. Sunflower seeds—Used as an oil for cooking or for making biodiesel, just to name two possible uses.

7. Canola—An oilseed, can be used as a high quality cooking oil or as a feed-stock for making biodiesel.

8. Soybeans—Another truly very versatile crop, because this oilseed can be used for livestock feed, human consumption in various forms, and as a feedstock for biodiesel.

9. Alfalfa—This perennial is a leguminous hay and is the Cadillac of forage to feed cattle in Kansas and in other states where the climate permits it to be grown. Top quality alfalfa sells at a premium to dairy farmers and is also an excellent crop for beef cattle and for cash income.

10. New field (biomass) crops—The intent is to grow grasses and other materials such as small trees to use as feedstock for a number of processes that can produce engine fuel and other energy related substances. Therefore the Fort Hickok community will also be serving as a prototype for some alternate energy processes that may now be in their infancy. These crops have not been grown to any extent on the Plains in past years as revenue crops.

    a. Switch grass—a tall grass that is used as a feedstock for biofuel.

    b. Miscanthus—also a tall perennial plant that can be used as a feedstock for biofuel.

    c. Spicy Mustard Seed—an oilseed that does well on marginal lands with minimal water requirements—another feedstock for biodiesel.

Chapter 2 Truck Garden Crops

Technically speaking, truck gardening (truck farming) is a form of field agriculture, and the crops listed below are to be grown primarily for human consumption although some could be used for livestock feed, particularly the spoiled items or those not suitable for marketing to people. Strange as it may seem, I know a lady that used to make her own diet formula for the greyhounds she raised commercially, and part of this Gawd Awful mixture was raw tomatoes. Yep, she mixed it up with her hands. Hogs would dearly love to lap up some taters, maters, and strawberries, and the franks, chops, bacon, and hams are excellent sources of cholesterol! What a deal. *Great* diet foods. Huh?

1.   Tomatoes—grown outside when weather permits.

2.   Potatoes—grown in fields for human food as well as for ethanol production.

3.   Green Peppers—grown outside when weather permits.

4.   Onions—also a seasonal field crop.

5.   Rhubarb—a winter hardy perennial plant used for pies, jams, etc.

6.   Asparagus—a high value perennial plant, hardy and producing when the weather permits. Some people consider it to be a delicacy. Oh my, different strokes for different folks, but I prefer horseradish?

7.   Strawberries—grown outside when weather permits, and are a perennial plant, at least for a few years.

Chapter 3 Greenhouse Products

These items bring a premium price in the colder months, the off season, and are usually shipped for hundreds of miles to reach the grocers in central Kansas. Local greenhouse production will help round out the supply of these products in the off season months when they cannot be grown in the fields.

1.   Tomatoes—used for eating raw, for sauces and juices

2.   Bell Peppers—an expensive item these days, even in the growing season.

3.   Flower sets—a favorite of home gardeners, particularly the gals.

4.   Bedding plants—both flower and vegetable species. Some are shipped hundreds of miles and they look it when they get there.

Chapter 4 Agribusiness Industries

These are primarily businesses that process cereal grains and truck garden crops. One might consider the making of fuel pellets and firewood to be an ag related business if farmers are to grow energy crops for pellets. Bless their hearts, if the farmers can feed the nation then they can also fuel the nation if we give them a chance.

1.   Flour Mill

2.   Cornmeal Mill

3.   Oat Milling

4.   Creamery or milk processing plant—for milk, cheese, cream, and other dairy products.

5.   Egg supplier—to produce and market farm fresh eggs (*cackle berries*)

6.   Bakery—breads, pastries, and other goodies

7.   Fish Farm—a newcomer to the barnyard set? Ever see a fish with legs?

8.   Cattle and Hog feedlots—to be slaughtered and processed outside the community, but nearby to conserve motor fuel.

Chapter 5 People Oriented Businesses

Some of the following activities are of a service nature and others are retail in nature, or are intended to cater to the tourist and recreation trade. Others are for local consumers or for whoever might drop in. Some are to provide for basic care for the indigent and the frail.

1.   Child Daycare—primarily for the local workers, but open to the public.

2.   Adult daycare—an alternative to living full time in a rest home, which is becoming cost prohibitive for many, particularly those who cannot qualify for government assistance.

3. Assisted Living—modest accommodations for modest budgets.

4. Boarding House—for those not wanting to do their own cooking and cleaning but in need of affordable housing.

5. Pet Boarding and Grooming Kennel

6. RV Park

7. Micro brewery—a pub with brews, music, and food

8. Motel—for tourists and others

9. Picnic ground—for families, groups and individuals, with paddle boats, adult games, boating, swimming, carpet golf, driving range, bumper cars, bumper boats, pontoon boats, excursion boats on the lake, fishing.

10. Car detailing

11. Dance club—snacks, games, microbrews

12. Liquor Store

13. Gift Shop and Bookstore

14. General Store

15. Country Opry—a theater that features live country music

16. Catering service

17. Healthcare Services

18. Dude Ranch—for families, for lovers. Great shades of Peyton Place!

Chapter 6 Miscellaneous Businesses

If all else fails then call it miscellaneous? Where are the cats and dogs in the mixture?

1. Fruit trees, berry bushes—The fruits are processed or sold fresh to customers. Some places have a u-pick-it operation.

2. Community vegetable gardens—primarily for residents of the community but space could be leased to outsiders if available.

3. Farmer's market—locally produced products

4. Flea Market—with emphasis on recreation, entertainment, snacks, dancing, lounging, and lotsa junkola (treasures!)

5. Cordwood sales and fuel pellet sales.

6. Shuttle service—locally and to nearby towns

Chapter 7 Alternate Energy Devices

These activities are an integral part of the design and the community effort to be more self sufficient and to minimize the impact on the environment. The rather comprehensive assortment of business activities in chapter 5 should be a clue that there is life after the oil patch and total dependence upon utility companies, and hopefully alternate energy can be used for those businesses.

1. Windmills for pumping water

2. Wind generators for electrical power

3. Hydrogen gas generators

4. Methane gas generators

5. Wood stoves (cordwood) for space heating

6. Pellet stoves for space heating

Chapter 8 Affordable Housing

No no, not junk housing, but that which is sized right built, and equipped for economical operation, maintenance and living.

Chapter 9 Healthcare Services

Health care ranges from minor emergency and routine cases to hospital procedures and stays. People drive for miles in rural America to see a doctor or to have a routine operation. Much of this could be done right in their home town if the staff and facilities exist there.

Chapter 10 Communal Living Concept

Communal living at Fort Hickok would more properly be called intentional community living. The intent is to provide the benefits of communal living without the rigors of a strict doctrine or code of conduct, and it is primarily an economic cooperative.

The Devil Is In the Details

Why the name *Fort Hickok*? How does an old military outpost, be it real or imagined, relate to the designed intent of the community? Frontier military forts were essentially self-sufficient when it came to providing for the needs of the troopers and their families, and people got by without a drop of motor fuel, but did need some kerosene for their oil lamps. I like to visit old forts and to see how these folks lived back then, and a local situation caused me to gin up Fort Hickok.

Abilene, Kansas, my present home town, is the place where Marshal Wild Bill Hickok worked for a short time, and there is a small amusement park on the south side of town that someone has been trying to rejuvenate. This effort has been in progress for a few years now with little to show for, and in the beginning of the renovation project my intuition told me that if the amusement park were to be operated as a museum it would probably flounder and fail for lack of patrons.

As I predicted, after five years it hasn't gotten off the ground to the point that it is a viable tourist attraction. Early on I had decided to design an alternative to the Old Abilene Town, and laid out and designed a fictitious Fort Hickok, which was actually a planned multi use retail development with the historical theme of a frontier military fort. The design concept was such that in the busy season it could serve as a tourist attraction and in the quiet season it could still serve as a needed shopping center for the local people. That in a nutshell is how my little Fort Hickok came to be. The date on the overall layout of the fort is on my birthday, October 25, 2002, and I must say it is pretty clever for an old country boy with two years of architecture training.

I mentioned my idea to the fellow who wanted to rejuvenate Old Abilene Town but he apparently had something else in mind so I simply put the design away in my field of dreams, which are many drawings of houses, buildings, developments, and cobwebs. So why did the revised *old town* fail? Wild west themes are passe by now and not in favor since it is no longer acceptable to have drunken fights, shootouts, and to pick on the Native Americans. Additionally, local color museums are a dime a dozen as each little town along the interstate highways seems to be trying to make a few bucks showing their heritage and history. Old Abilene days must have been very much like the TV series *Gunsmoke*, and as much as I love to watch it, those types of programs have fallen from favor, long ago.

I like to visit historical sites once in a while, but one visit is enough if there is nothing to draw me back again, so I have chosen to use a historical theme to increase the appeal of what is essentially a working community with stores, industries, recreation services, and housing. Look for a more detailed description of my original Fort Hickok in the Appendix. It covered two city blocks, which was about two thirds the area of Old Abilene Town, but the Fort Hickok in this book is much larger than that and is much more comprehensive in scope.

I once studied architecture, and then spent my adult career designing and coding computer programs, so being a creative sort of person, I tend to collect such little scraps of paper, which are part of my precious memories. Additionally, I am an amateur photographer and have a good 35mm SLR camera that records my travels to the attractions such as old forts and ethnic villages. I wish it were possible to share my albums with you because the heavy Pentax does a superb job and is smarter than I would ever be when it comes to taking good pictures. I am primarily interested in the architecture and the rudimentary ways that people used to make their way thru life back then, and the architecture is a lasting tribute to their venture.

I became interested in alternate energy forms and methods in 1973 when America experienced one of the earlier wake up calls to the hazards of being overly dependent upon imported oil and overly dependent upon a sole source of motor fuel, that being petroleum. In the pages that follow I hope to present what I think are some alternatives to hydrocarbon fuels such as coal, petroleum, and natural gas, and I would like to incorporate these ideas into my new Fort Hickok.

Some of the new ways to consume alternate energy differ little from the present ways of using petroleum based fuels, and some require a new approach to best use. It is the new approaches that may spell the success of using alternate fuels more than the fuels themselves, because even alternate fuels are finite in supply and our appetite for motor fuels seems to be insatiable. And the surprising thing is that in many cases the best approach is to simply do less to get the same results. It is possible, as you will see.

1

Traditional and New Field Crops

1. Wheat

Wheat is a mainstay of Kansas and of many Plains states and, is a relatively high value crop as cereal grains go. So why include this in the assortment of crops? First of all, it is a great cash crop since wheat requires little attention from the time it has been planted until harvested, and it commands a high price per bushel. It is also hardy enough to do well with the harsh climate of Kansas and does not require as much input (especially water) as some of the other crops I will be discussing.

There was a time when large flour mills were within a 100 miles of most Kansas wheat fields but consolidation has changed that. Only a few specialty mills remain in the interior of the state at cities like Salina, Hutchinson, and Wichita, so the bulk of the mountains of wheat grown here each year has to be transported, usually by train, to points east such as Kansas City, St. Louis, Chicago, and other long time milling centers in the more populous states of the north and east. When you add to this the fact that one third to one half of the annual crop is exported you can see where this is a very fuelish activity and one that I don't think is sustainable from a domestic energy standpoint. Wheat exports typically leave the Great Lakes area and go to the Atlantic Ocean if eastbound.

By now you might be wondering where Kansans get their flour for home use, and where the flour for store bought bread and other baked goods comes from. It is shipped back westward from the eastern milling centers at the expense of huge quantities of motor fuel to propel the large trucks and the husky railroad engines, a whopping 4,300 hp per locomotive, a marvel of engineering that gulps diesel fuel at the rate of *4 gallons per mile* when it is pulling in earnest. If you don't find this practice to be wrong headed and wasteful then ask yourself where your flour comes from and why. Don't we have more critical uses for scarce diesel fuel?

The relatively small consumption of wheat by the flour mill in Fort Hickok would be insignificant compared to the whole supply of wheat and flour that is moved across the continental United States, but the principle is valid. If your bread costs an arm an a leg and the flour prices are out of reach, a major factor in the price of our food these days goes to diesel fuel needed to haul the products in their various stages of manufacture or production. Hickok's people will simply try to do the right thing, wanting to live within their means in the energy sense. "Waste not, want not." they say. By also cutting out the middle men I think considerable savings can be realized in baked goods and flour. This again is another tenet of the virtue of simplicity and can be applied to many economic endeavors. Cut out the waste and do more with less motor fuel consumption.

Just a word about the use of wheat straw as a feed stock for biofuels. In recent times new varieties of wheat have been developed that have a shorter stalk, which is the straw. This development is thought to be an improvement over the older varieties that grew tall and heavy, maybe waist high in the rich river bottom soils, which was a sight to behold. I once had such a field of wheat out back of the house and my cocker, Ladybug, went snooping in the wheat and got lost. She couldn't jump high enough to see the way out so a loving daddy rescued her from her predicament by calling her as she came to him. That straw that they have engineered out of the wheat field could be used as a secondary product to be sold by the farmer to a biomass processing factory that makes fuel from it. So much for the march of technology. Maybe some of the old strains can be resurrected so we can have both grain and straw from our efforts? The partial oxidation process can be used to convert wheat straw to methanol, and it will be discussed in Appendix D. Cellulosic ethanol can also be made from wheat straw, and this process is discussed in Appendix D.

The example of local grain milling illustrates one of the primary ways to gain leverage in the quest to conserve fuel and to have an affordable food supply. Hey guys and gals, here is the silver bullet, quit hauling the darned stuff half across the country and back. As simplistic as that may seem, can you fault the reasoning when it comes to a low cost, low tech, and sure fire way of conserving motor fuel? As I continue the same principle will be applicable in other endeavors of the Fort, a simple and straightforward method that is fail-safe and has stood the test of time. As our domestic supply of petroleum dwindles this basic premise of conservation will continue to have value in the coming years, so in this respect the concept is on the right track and may be our saving grace, our panacea. If the corporate barons could only see it? It is not for sale but it's there for the asking, if they could only see its power and value

The key element in successful centralization of business activities is a fuel supply that is both affordable and dependably available, and those traits of our fuel supply are coming under increasing pressure as time flies by. The average consumer frets about the price of gasoline and I can sure relate to that, but there are a few things he may not know about diesel fuel. Diesel is the choice of fuels for heavy engines that develop large amounts of horsepower because a gallon of diesel carries more energy than a gallon of gasoline or ethanol. The down side of this is that there are fewer ways to make a biomass form of diesel, or biodiesel, than there are to make a substitute for petroleum based gasoline.

One more tip on the habit of using diesel fuel rather than other fuels. Yields of refining processes can vary, but in general a barrel of crude oil yields about half as much diesel fuel as it does gasoline. As long as crude oil is being refined for the diesel users the refiners are going to want to sell the gasoline too. That essentially makes the gas burning motorist hostage to the diesel users, and the oil patch barons will do what they can to keep it that way. Fort Hickok and its concepts are intended to alleviate that pressure and to assure people a dependable and sustainable way of life. We don't have to stay over the barrel, particularly the crude oil barrel.

There are other reasons for the community growing its own wheat. The type and quality of the grain can be controlled, and the safety of it can be better controlled than when importing wheat gluten laced with Melamine, just to save a few bucks but not caring about man's best little friends, our cats and dogs. Even farm raised fish may have been fed the Melamine powder, which is used to make plastic dishes and such.

Kansas wheat is typically red winter wheat, or hard wheat, and it is good for baking bread and pastries. However, a variety such as Durum makes better flour or semolina for pasta, and this too is something the community mill could produce if it were in the right geographical location. The online Wikipedia encyclopedia has an article on durum wheat and states that most of the durum produced in America is grown in North Dakota, approximately 59% of the United States crop in 2004. What the heck, if you cannot grow durum in your area then buy it, mill it, and still avoid the expenditure of fuel to haul it back east to a mill and then haul the finished product back home to be consumed.

An additional benefit for growing, milling, and consuming a product in somewhat the same locale can be seen from looking at the composition of a grain of wheat. It has a skin or hull, a germ, and the starchy substance called the endosperm. The milling process separates these components and bags the refined endosperm as flour, bags the hull as animal feed, and sells the germ for other uses. The germ tends to be oily and if left in the flour it would not keep well.

So why ship wheat several hundred miles to mill it, only to ship the hull, called bran, back to the farm to feed cattle? If you want to see stock cattle then come to Kansas because those big fuzzy creatures are everywhere and the tonnage of meat that is shipped from Kansas is staggering, also a tragic waste of valuable diesel fuel, relatively speaking.

A grain of wheat consists of the following parts: (1)
    83% endosperm, which is the flour
    14% bran, which is the outer layers of the grain of wheat, the cow feed
    3% germ, which is sold as a dietary supplement or as cow feed

From my experience as a farm kid of a family that grew wheat, the maximum level of moisture in the grain that is safe to store without heating and spoiling is on the order of 12%, so for every ton of wheat being shipped (2,000 lb.) there is 240 pounds of water. Yes, water is an essential part of the grain, but must we haul water half across the country and back? Surely by now you can see the potential savings of diesel fuel if we simply grow, mill, and consume our wheat closer to the source and closer to the consumer. Somewhat the same reasoning can be applied to the other cereal grains I will discuss. Appendix H has information about the milling of wheat and K lists some of the existing flour mills in Kansas.

I think that the capacity of the flour mill at the Fort can be much larger than just enough to serve the local population, which I estimate to be 6,000 people once the town has matured. If by producing wheat flour locally and then shipping some of it out to other areas, or to a nearby major population center, then the principle of fuel conservation can still be observed because the distance the flour would be hauled would still be much shorter than the hundreds of miles it may come from a major mill in a larger town.

2. Oats

The tonnage of oats that Americans consume pales beside the tonnage of wheat and corn, but oats is a grain with special qualities that make it desirable as human food and for feeding livestock. Horses especially like oats. Hey, I am an old duffer now, and I gotta have my oatmeal to keep me regular and not so cranky, if you know what I mean. So if eating oatmeal cereal is like taking medicine then eating oatmeal cookies with nuts and raisin in them is like being a hog in Heaven. Yum. Yum.

Oatmeal comes from the meaty part of the grain, the inner kernel, which is called groat in the milling trade. I liken it to the endosperm of the wheat grain. The groat is rolled flat to become flaked oats (oat meal) in its dry form. To sum it up, let's grow our own oats and make our own oatmeal and other oat products.

A bushel of oats doesn't bring much compared to wheat and the yield per acre usually isn't high so growing the product is not attractive to most farmers. The relatively low price per bushel of oats tends to concentrate this activity in certain locales, and this concentration brings on the nagging waste of fuel needed to haul the stuff half across the country to be processed and then back to the consumers. Oat grain is bulky and weighs only 32 pounds per bushel standard weight, so a big truck can be tooling along the highway with a full load of oats and still not be carrying much tonnage. If the truck gets little better mileage running light than it does hauling a heavier grain such as wheat then the haul is even less cost effective in terms of diesel used per ton of product. This silliness reminds me of hauling potato chips, a lot of air and paper but not much potato. The modern milling process for oats is clever and is a departure from the milling process used for wheat, so look for more details in Appendix J.

3. Barley

As with oats, the amount of barley grown is quite small compared to wheat, but it serves a purpose in beers, whiskeys, soups, and in other foodstuffs. It too tends to be a specialized crop grown primarily in certain areas, which again brings upon the waste of fuel associated with hauling it around the country in various forms. According to the Wikipedia encyclopedia (2) barley is 15% water by weight, so once again, we do not need to be hauling water any further than necessary. As a side note, the United States produces a relatively small amount of barley where Russia produces a very large amount, comparatively speaking. Vodka anyone?

A grain of barley contains a hull, much like oats does, and this part is inedible for humans. The remaining parts are the groat which includes the endosperm and the skin, or bran as in wheat. I see this as being a grain that could be milled and used to produce specialty food items as well as beer, soup, and breads. Sometimes health food stores want barley flour in lieu of wheat flour for their customers, so that is even more potential in this grain. Let's avoid shipping the finished product all over Creation and back because to do so would be hypocritical.

4. Corn

Corn is a versatile cereal grain and is used to feed livestock, to make various human foods, and as a feedstock for ethanol. The primary intent of growing corn in the Hickok community is to have it for animal feed and to make cornmeal. The Wikipedia encyclopedia has an excellent article on corn under the name of maize. From my experience the moisture content and the various edible parts of the grain are similar to that of wheat.

The United States is the major producer of corn in the world, and slightly more corn has been raised in recent years than wheat. However, with the new interest in making ethanol from corn you can expect the acreage of corn in the United States to increase dramatically and the price per bushel to also increase because of the additional demand for the grain. I find this to be a bit fuelish and short sighted as corn requires a great deal of input in the way of fertilizer, weed control, and water to produce a high yield, and it won't do well in some of the drier Plains states because of the heat, so the long hauls go on in the name of making a green motor fuel. Guess what fuels the big trucks and trains for the most part, petroleum based diesel fuel. Hmmm?

One of the drawbacks of producing more ethanol than can be consumed within a short distance is the problem of transportation. Apparently ethanol cannot share the same liquid pipeline with gas and diesel fuel so it must be hauled by truck or by train tank car. So here we go again, the old bogeyman of centralized production of ethanol in a relatively few number of plants and the tremendous expenditure of diesel fuel needed to transport the corn from the farm to the ethanol plants, and even more scarce diesel fuel to haul the ethanol to the blending plants, which are usually oil refineries.

An additional consideration is that once we devote too much corn to ethanol then the other users of corn will bid up the price to get enough to make their products. Given the fact that we can never make more than a small percent of our fuel from corn based ethanol, I think we are kidding ourselves. There are other ways to make motor fuel, which I will explain upon later in the book, and there are alternate feedstocks for making ethanol.

The purpose of raising field corn in the Fort Hickok community would therefore to save the waste of fuel needed to haul the stuff around. Let's grow it, process it, and use the resultant products right here at home, and let's resist the urge to ship large quantities out of the area for profit although that may be the best way to dispose of excess quantities. A herd of hungry hogs would be a less fuelish alternative as little pigs can be easily had and grow to slaughter weight in a few short months. Personally I love my franks, sausage, ham, and chops, but I can't stand the thought of killing the playful little critters. Appendix I has information about milling corn.

5. Milo or sorghum grain

Milo is an alternative to corn in areas where there is insufficient rainfall or irrigation water for a good yield of corn. It is relatively easy to grow and harvest, and it makes a good feed for livestock. The Fort Hickok community would grow it primarily for animal feed. Wikipedia tells me that it is roughly equivalent to corn in food value as an animal feed. Some cultures use milo flour or use the cracked grain as a porridge but Americans in general do not do so. An alternate name for milo is grain sorghum, and the best I can remember milo was introduced into Kansas after WW II as a short stalked sorghum grain that could easily be harvested with an ordinary wheat combine or harvester. Harvesting it is an itchy job but the grain is easy to work with and makes a nice livestock feed, especially for hogs and cattle. The grain is a small round berry about the size of a peppercorn, and can be brownish in color or an off white. It is kinda fun for a hungry farm boy to chew on too.

6. Sunflower seeds

These seeds are grown primarily for oil, which can be used for cooking and to make biodiesel. Sunflowers do well in Kansas and would make a good cash crop, and a better one for biofuel than wheat as wheat has not been used to any extent to make ethanol to date because of its relatively high cost per bushel as a feedstock. Sunflower seeds would be crushed and the oil used to make biodiesel or cooking oil. I see it as a cash crop for a community if the distance to the crusher is short because there are relatively few varieties of seeds or grains that can economically be used as a feedstock for making biodiesel.

7. Canola

Canola seed is an excellent source of oil for cooking and for making biodiesel, and once again this product could be a cash crop for a community such as Fort Hickok. Selling biofuel crops for cash income is a way to contribute to the effort of having a renewable and environmentally friendly fuel supply. To date there has been relatively little experience in the central Plains states with canola but it has been grown successfully in the northern Plains states and extensively in Canada. In time you will see why I think the Fort Hickok community will need some income from cash crops and sales to outsiders of its products. The income will be needed for the people oriented plan. Keep in mind that the business organization of the community is essentially one or more diversified corporations. In other words it is a community of businesses as well as a land area having people.

8. Soybeans

This little critter has been grown for centuries and is used for animal feed, for human consumption, and can be used as a feedstock for biodiesel although it is a pricey feedstock because of the terrific demand for soybean products across the globe. An oil press for soybeans is commonplace in the Plains area, so the technology is mature and I think one could have a small scale plant if desired. Short of that, soybeans would be a good cash crop and an ecofriendly addition to our feedstock for biodiesel. Hopefully a crusher or oil extraction facility could be fitted to process more than one kind of oilseed, such as sunflower seeds, mustard seed, and canola (rape seed) in addition to soybeans.

9. Alfalfa

Alfalfa is a perennial legume, a hay for cattle, and is an excellent forage for meat and milk cows. Alfalfa commands a premium price if the quality of the hay produced is top end, so it can also be a good cash crop. Dairy farms and feedlots pay dearly for top quality alfalfa hay, but its bulkiness makes it a rather fuelish proposition to haul it long distances to market. It is also a soil builder because it somehow captures nitrogen, where other crops simply consume what is available. An alfalfa field can stay productive for as long as 5 years, possibly more, if the crop is taken care of by adding lime as needed and by controlling the weeds and the grasses that love to compete with it. The longevity of the plant plus the fact that multiple cuttings can be taken from the field in a normal rainfall year makes it a good dry land cash crop. However, alfalfa will not tolerate standing water, so the field needs to drain well and to be out of the flood plain.

Horses also love alfalfa but one must be very careful about feeding it to them as there is an insect that can end up being baled in the hay, and if a horse eats even a small part of that insect it can become very sick and even die. I think this pesky varmit is more apt to appear in the first cutting of the year, which is the premium cutting in terms of tenderness and protein. That killer may be called a sow bug but don't quote me on it. Thank Goodness it does not affect cattle or hogs.

The equipment needed to efficiently grow, harvest, and handle alfalfa bales has become sophisticated and expensive, but most of the hard manual labor has been taken out of alfalfa production. This is a beautiful crop that can benefit from generous visits by the honey bee and it makes a good source of pollen for the bee to convert to honey because there are jillions of little blossoms on maturing alfalfa hay. Additionally, honey is a local alternative sweetener and perhaps a healthier substance than imported refined sugar.

Abilene, Kansas has a dehydration mill which takes the cut alfalfa, dries it, grinds it, and finally extrudes alfalfa pellets, which are a convenient way to store, transport, and feed this very important product. You may also have seen alfalfa as rabbit food in a pet store, little cylindrical greenish pellets about an eighth inch in diameter and about a quarter inch or so long. The dehydration process can produce a strong odor, so hope you are upwind from it, but alfalfa in the field has a nice fresh smell. And the milk and dairy cows go ape over it. They never seem to outgrow their taste for it, that is cured alfalfa, because raw or fresh alfalfa can cause an animal to bloat and die if that is all the animal has eaten.

10. New Field Crops

These are hopefully the major fuels crops of tomorrow, and since I have had no experience with them I may seem to be as lost as you are when it comes to discussing them. Let's shorten the agony by first saying that these crops are to be used as biomass input to processes that can produce biofuels of various kinds. Later I hope to touch upon the processes that I think could be utilized in the future to make biofuels.

Consider this possibility for your locale. There are millions of acres of marginal soil present in the Conservation Reserve Program (CRP) and people are paid an annual sum by the federal government for keeping the land out of production. Some sort of cover crop is mandated in order for the land to be in the program and it is often a tall coarse grass. If a portion of this CRP acreage could be taken out of the program and used to grow biomass for fuel crops then the taxpayer would get a break from coughing up the billions of dollars each year being spent on the program and we could have another renewable alternative to petroleum based fuels.

A valuable and unseen virtue of biomass crops is that they are renewable, year after year, as opposed to a field of oil or a vein of coal petering out after relatively few years of bounteous production. One of the reasons that coal and petroleum products are so expensive to American consumers is that the easy pickings have been worked out, but we can avoid this inevitable trait of hydrocarbon deposits that lie underground by going green and renewable. Another virtue of a biomass crop is that you don't have to explore for it, you simply grow the darned stuff, so you will surely know where it is, unless you have senior moments like I do, and then you might be excused.

a. Switch grass

This is a tall perennial grass that can be burned directly after drying and cutting into small pieces or it can be made into fuel pellets in about the same manner used for making alfalfa pellets. The heating value of switch grass compares favorably with wood and with wheat straw. Yields do not mature until the third year after seeding, and are in the range of a ton or more per acre, depending upon the soil being used. It can be used as forage for cattle but I am not aware of this being practiced to any extent in the United States. Perhaps not feeding it is a plus because it tells me that other uses are not competing for it with the fuel pellet processes. Compare that to the situation we face when corn for ethanol competes with corn for other economic activities, not the best of predicaments. An additional virtue of switch grass as a feed stock for pellet fuel is that it burns very completely, about 3-4% of original mass remaining as ash (4). I see this as an alternative feed stock to oak wood for fuel pellets in the locales where large amounts of oak is not available, which is the case in parts of Kansas, and the crop can be harvested once a year where it may take a decade or so to grow a nice sized oak tree that is feasible to cut for cordwood or used for fuel pellets.

Switch grass has an excellent ability to sequester carbon in the soil on which it grows, and this is considered to be an important strategy for reducing atmospheric CO_2 (5). In time I will present some relatively new processes than can be used to convert biomass such as switch grass to a type of motor fuel plus other useful products. Some of the processes do not require fermentation of seed grains or starch derived from cellulose to produce fuel.

b. Miscanthus (6)

To date (May 2007) there has been little experience in the United States with this tall grass but parts of Europe have been working with it. The mature plant can reach heights on the order of 11.5 ft and the growth is very dense, which means high tonnage per acre of this crop. Once again, it can be used as a bioenergy crop and it is harvested with a sugar cane harvester. Miscanthus can be grown in the cooler climates and can be converted to ethanol, so this may serve to help the strain or demand on our corn production acreages.

The crop is perennial and is started by planting rhizomes, which are pieces of the root, and harvest can be started the second year. In time the plants will spread to become a very thick stand of tall cane like things that resemble hybrid sorghum, which can be grown as a hay grazer for cattle forage. I also liken a stand of Miscanthus canes to bamboo poles! Virtues of the plant as a bioenergy source are relatively high yields, on the order of 3-6 tons per acre, annual harvest, and low mineral content, which improves fuel quality. Given the latter trait, Miscanthus might be a good candidate for input to fuel pellet production. I will say more later about my experience with my pellet stove, and it has been a good one so far.

c. Spicy Mustard Seed

The spicy mustard seed is a good candidate for input to a biodiesel process and the plant can give acceptable yields on marginal soils that aren't much good for other crops. Mustard meal is also a high value pesticide for the organic market. So once again, here is an alternative to the old grind of the usual field crops and the petroleum based diesel fuel. This is an annual crop that is planted from seed, and a mustard seed is about 40% oil where a soybean is only 18% oil, so in that respect it will be a good competitor to the traditional oilseeds.

The yield in pounds per acre seems low relative to that of other oilseeds, about 480 to 850 pounds, but apparently folks in Colorado (3) are attempting to develop the practice to a commercial scale. The name of the game is to have an affordable feedstock for making biodiesel. If a farm cooperative could make biodiesel from oil seed then it seems the farmers would be more than happy to use it if the price was right and the quality was there.

I could go on and on about various biomass products that could be used as feedstock for processes that make motor fuels but the scope of this work does not permit that. I do plan to illustrate how versatile grasses, crop residues, and other organic materials can be used as feed stocks to biofuel production, but that in a sense is a bit outside the scope of this book, so I will simply offer a brief outline of it as justification for producing some of the crops I have listed. In general, green fuel is made from either the grain or oil seeds of plants, or from the cellulosic material of the leaves and stems.

For example, I once read in a Department of Energy online publication (3) that there is enough biomass to create a huge amount of methanol, which can be used for fuel much like ethanol can except that methanol is caustic and harder to work with than ethanol. The projected annual output is more than enough to replace our current usage of gasoline, and the beauty of it is that the feed stocks are renewable. Methanol can be made from trees, shrubs and other woody plants, and is sometimes called wood alcohol. I think we are missing a golden opportunity in not developing stationary applications that could use methanol, and I say stationary because those systems could include the provisions needed to tolerate the use of methanol.

So once again here is a challenge to make do with what we have available domestically rather than being dependent on imported oil and fuels. Rather than driving our own gas guzzling cars we could use the methanol fuel to power an electric tram or could even feed it to a specially designed fuel cell that would generate electrical power. I'll bet those Amana pioneers would have been using it if the technology was known in their era.

Let's move on to something that most people can easily relate to, fruits and vegetables grown truck garden style in fields. At last, as grass doesn't taste very good to me. I am a mouse and not a rabbit.

2

Truck Garden Crops

Fruits and vegetables contain a high percent of water, as much as 75% or more and the stores are stocked with products that have been shipped as far as two thousand miles from the producer to the local grocer. How much do you want to pay for your water? Granted, the water content of a fresh tomato or a potato is essential to maintain the quality of that product being fresh, but must we be held hostage to the high transportation costs that bring us those things? Most fruit and veggies are carried by tractor trailers (18 wheelers), and these monsters guzzle diesel fuel at the rate of five to eight miles per gallon, so the cost of freight, which fuel is a part of, is a significant part of the cost of a produce item if it has been shipped a long distance.

The next time you buy tomatoes or potatoes check out the origin. Here in Abilene, Kansas the stores carry tomatoes from British Columbia and potatoes from Idaho. Both locations are a long way from central Kansas, on the order of 1200 to 2100 miles one way. To make matters worse, potatoes and tomatoes are very high in water content, and you cannot get nutrients from that water so you might as well get it from your tap. It is a matter of getting the best dollar value for our fuel expenditure, and that value can be greatly enhanced by growing many fruits and vegetables close to home.

Mapquest (1) tells me it is 2027 road miles from Vancouver, B.C. to Kansas City, Missouri (KCMO), which would be the likely distribution center for tomatoes being shipped into the Abilene area, so we are looking at roughly 2160 miles from the tomato producer to the local grocery store. In the case of apples, it is 1718 miles from Wapato, Washington to KCMO, and another 150 back to Abilene, making a total haul of 1768 miles for apples coming in from Washington state. In the case of potatoes coming from Boise, Idaho you are looking at 1368 road miles to KCMO, and another 150 miles back to Abilene, for a total haul of 1518 miles. Assuming a diesel powered truck (18 wheeler) gets 6 mpg, one trip of 1768 miles would use 253 gallons of fuel.

Now 253 gallons of diesel may not seem to be significant, but if one were to estimate the total fuel bill for hauling just these three food items to little Abilene, Kansas, which is about 6,000 people, then it would be a sizeable number of gallons. The cost of this fuel is part of the cost of freight, which becomes part of the cost to you the consumer when you buy these things from the store.

The details tell the story, but the thing to realize is that by growing these popular products nearby diesel fuel could be saved for the more critical applications that need it, like the heavy coal trains, the earth moving equipment, and the fire trucks. Short hauls can be economically made in lighter trucks (with smaller loads) that can burn a type of biofuel other than diesel, which is another point I was wanting to make. There is an alternative to petroleum based diesel fuel and biodiesel called syngas, whick will be mentioned in Appendix D, Gasification.

Let me also include some facts about the moisture content of various foods in order to avoid the tedious job of referring back to the Appendix Tables for the information. These moisture contents of selected food items were taken from a web site on the internet, and I will include a larger list in the appendix. The following water percentages were taken from a chart by Walton Feed, Inc. (2) and the per capita consumption was taken from Appendix E, Table 2. The data can also be used to calculate how much of each item is consumed at a given locale, knowing the population of that area. See Appendix E for Table 2.

Item	Percent Water	Per capita consumption per year
Potato	85%	46.2 lb. farm weight
Red Ripe Tomato	93%	17.4 lb. farm weight
Apple	85%	15.5 lb. farm weight
Carrot	88%	10.6 lb. farm weight

Doesn't it strike you that hauling tomatoes over 2100 miles from the farm to the consumer is a bit wasteful and should be avoided? The water content of processed foods is surprisingly high, and I say that it is unnecessary and fuelish to continue such practices.

A few of the truck garden products can be grown in greenhouses in the cooler months but this section is concerned with commercially growing these items outside in fields. Yields per acre will vary depending on the climate and length of growing season. Refer to Appendix E Table 4 for the yield of selected food crops. Crop years may vary depending upon the source of the data and I sometimes use a figure that is representative of an average yield rather than one that may be low or high in a given year. The intent of showing these numbers is to give a general feel of how much can be produced per acre and it is not to be held to a precise number. Also keep in mind that a crop grown in California, for example, may yield much better in a year than a northern state where the growing season is shorter than that in California.

1. Tomatoes grow well outside in the growing season, and locally produced vine ripe tomatoes are much better tasting than those that are shipped green from a distant greenhouse or field. The average yield per acre of tomatoes is 29,900 lb. (3)

2. Potatoes are normally grown in fields and the yields per acre are fantastic, on the order of 38,200 lb. per acre (3) per season. Given the high consumption per capita of potatoes, significant savings of transport fuel can be had by growing them locally or at least within a 100 mile radius of the consumer. I use the figure of 100 miles because the land mass of Kansas can be divided into four essentially rectangular quadrants, each being 200 miles wide and 100 miles high. So a strategically located potato patch within a given quadrant would have a maximum haul of 100 miles from the farm to the most distant consumer in that quadrant. This rough estimate is intended to illustrate the concept of conserving fuel by keeping the production of products close to the consumers, and the principle can be applied to a variety of products, even services for that matter, as you will see me illustrate with the recreation and travel activities I will be discussing.

3. Green Peppers can be grown in fields and are also a high value crop for a greenhouse environment in the off season. Do they ever get cheap these days, in season or out of season? The average yield per acre is 28,400 lb. (3)

4. Onions are usually grown outside in fields and surprisingly also have a high water content at 89%. Prices for onion have sky rocketed in my locale in the past few months, and a decent white onion will now bring 98 cents per

pound or more where they used to sell for less than 70 cents per pound. The average yield per acre for onions is 44,000 lb. (3)

5.   Rhubarb is a specialty crop, a celery-like plant in texture with a tart taste, and is used for pies, jellies, and other delicacies. It is a perennial plant and a low maintenance plant, so I see it as an attractive sideline for a community that offers home style cooking to patrons. Rhubarb is seasonal, and is usually harvested early in the spring each year.

6.   Asparagus brings a high price and is a hardy perennial. Once again, a specialty to be offered to patrons of a nice restaurant in the Fort Hickok area. It is being offered for $2.48 a pound today (June 13, 2007) in the local market.

7.   Strawberries are a national favorite and although they require a considerable amount of care and input each year, they are perennial and are a high priced item in the stores. And there is nothing like raiding mom's strawberry patch before the birds get the big juicy ones, just ask any ornery little farm kid. More information about this tasty fruit can be found in Appendix N. The average yield per acre is 44,500 lb. (4)

3

Greenhouse Products

Greenhouse production is badly needed at the local level to provide people with items that are being imported from afar, particularly in the cold months, and the rewards can be good in spite of the work and expense that is involved. The trick is to offer a top quality product and to keep the energy costs low.

1. Tomatoes—A bargain these days for a tasteless tomato that has been shipped into the area from afar is $1.69 per lb. or more. The nicer ones sell for over $2.00 a pound.

An article titled *Green House Tomatoes* by Charles W. Marr of Kansas State University (1) gives the basics of sizing a greenhouse, the operating schedule, and the yield per square foot of greenhouse space. Four thousand square feet, or two houses, each 22 feet by 96 feet, is enough to supply the market for about 10,000 people. From the table in the article, given an annual yield of 2-3 lbs per square foot for the spring crop and 1-2 lbs per square foot for the fall crop, two greenhouses will produce from 8,000 lbs to 12,000 for the spring crop and 4,000 lbs to 8,000 lbs for the fall crop. In other words about 12,000 to 20,000 lbs will be produced per crop year (two seasons), which is only part of the total supply needed for a city of 6,000 people, the size of Fort Hickok.

Harvest of the spring crop of greenhouse tomatoes would run from April 15-30 and would end from June 15-July 1, and harvest of fall crop greenhouse would run from November 10-15 and would end from December 25-31. Figuring three months of production for the spring crop and about six weeks production for the fall crop, we would have production about four and one half months, or approximately one third of the year. If the demand for one third of a year (for 6,000 people) is 39,600 pounds, using the data from Table 2 of Appendix E, then as many as three of the 2,112 square foot greenhouses will be needed and no less than two will be needed.

If the total demand per year for 6,000 people is 118,800 pounds of fresh tomatoes then this leaves 79,200 pounds to be raised in the field during the growing season for the area, or possibly in additional greenhouse space. Table 4 allocates four acres to grow the entire demand per year (fresh tomatoes) for 6,000 people, so if that production were added to the annual greenhouse production there would be a surplus to use for processing into juices, canned tomatoes, and other tomato products.

Two more points before moving on to peppers. Greenhouse production of tomatoes, if a two crop system is used, calls for labor nearly all year, and only a month off during July will be offered because the workers will be kept busy starting the new plants and cleaning up the house for the next season. So this is a rather permanent part time job for two or more people, depending on the number of greenhouses being operated. The KSU article indicated a price per pound of $.85 to $1.30 per pound, which I take to be the wholesale price. If some of the product were sold at retail in a store that is part of the overall operation then the profits to the business could be enhanced, even without charging the exorbitant prices I have mentioned in the opening statements to this section.

2. Bell Peppers—Check the price of peppers in the store. Green bell peppers go for less than $1.00 each and the other colors sell for $2.00 a pound or more.

The following link provides a comprehensive look at producing bell peppers in greenhouses although it pertains to Florida production for the most part. See http://www.ars.usda.gov/is/np/mba/jun05/pepper.htm (2). A search for "green pepper production" using Google will yield many topics relative to greenhouse production of peppers, and one can quickly see that it is being done in the northern climates such as in Alberta, Canada and Oregon state as well as in the southern parts of the United States. Red, yellow, and orange bell peppers are very popular and bring a premium price over that of green bell peppers although the yield is less than for the green peppers.

Red and yellow apparently yield from 1.6 to 3.0 lbs per square foot of greenhouse growing area, and orange ones from 1.4 to 2.2 pounds per sq ft of greenhouse growing space. This subject has a lot of facets to consider if wanting to grow peppers in a greenhouse so I recommend reading the web site above and also the link cited in the article itself.

Table 4 of Appendix E shows a demand per year of 28,400 pounds per year for 6,000 people, so using a yield of 2.0 pounds per square foot for greenhouse production, then 14,200 square feet would be needed, or between three and four of the sized greenhouses illustrated above for growing tomatoes.

I think that it is sufficient to say that we no longer have to remain dependent upon the southern states for all of our off season tomatoes and peppers and can grow much of our needs at home. If half of the annual demand for 6,000 people (14,200 pounds) were to be hauled some 1,500 miles from Florida or California to central Kansas, it would take part of a semi truckload, perhaps half of one. At a fuel consumption rate of five miles per gallon it would take 300 gallons of diesel fuel per one way trip, and half a load could be charged with half that much fuel, or 150 gallons. This is not much for a mere 6,000 people, but when you consider the fuel needed to serve millions of inlanders then it becomes very significant. And tomatoes and peppers are merely a drop in the bucket when it comes to hauling fresh and processed produce around this country.

3. Flowers—usually in time for spring planting in the home garden

Potted flowers, ready to plant by the home gardener, often come hundreds of miles from the grower to the retail outlet. We can save fuel by becoming a grower at Fort Hickok, and if we have some excessive capacity then we can ship them short distances to retail outlets.

Each year the major supermarkets and the home improvement centers, even the big discount department stores, go all out to sell potted plants, particularly annual flowers. And each year the same customers come back for more annual flower plants. Why can't we grow our own right in the community by operating flower greenhouses too? It is a good job for those who like to work with potted plants, and it would be a part time job if only potted flowers were produced, so I think there would be plenty of help from folks of all ages to work steady or part time in these greenhouses.

4. Bedding plants—also a seasonal item, and apparently a lucrative one

Each year the local stores offer bedding plants, and I am speaking mainly of vegetable plants ready to set out in gardens. Fort Hickok will need such plants to support its commercial growing operations so why not produce a few more to be sold to individuals and save having them being shipped hundreds of miles into the area? Another virtue of locally grown plants is that they will most likely survive the local climate conditions.

Appendix E, Table 2, provides the per capita consumption of selected items, Table 1 provides the percent water of selected foods, and the capacity of truck farms needed to produce an amount to satisfy the demand in a given quadrant of Kansas will be provided by Table 4. The reader can do the same for an alternate state of residence. Different crops do well in different parts of the country, so this must be taken into consideration when designing a production program for a given locale. Greenhouse production rates also depend upon the locale and the length of the growing season.

4

Agribusiness Industries

These activities typically process farm output or are a specialized part of meat production such as a commercial feedlot. The primary objective of having these business as part of the Hickok community is to reduce the amount of fuel spent to transport the feed stock to the business, such as when hauling wheat to a flour mill, and to reduce the amount of fuel needed to get the finished product back to the consumer. An additional objective is to make some profit from these tried and tested ventures so the new ventures can be subsidized. For example, some processes for producing fuel from biomass may not be all that profitable until the technology matures, but pursuing such a venture is one way to learn a better way to do it. There is nothing like field testing to prove a process.

There are also activities of the Hickok community that are essentially not for profit but need funds to provide the benefits. Hopefully some of the profit from these ventures can be applied to those benefits, which are in the people oriented section that follows.

1. Flour mill
   If the economy of scale is such that some of the output must be shipped out of the area then that is a consideration. However, care should be taken not to defeat the principle of motor fuel conservation just for the sake of profits or having a local mill. See Appendix H for more information on wheat milling.

2. Corn Milling
   The same goes for cornmeal output as for flour output. Hopefully a small scale plant can be had to serve the local needs. See Appendix I for Corn Milling

3. Oat Milling

Milling of oats is a different process from that used for wheat and corn, and I found it to be pretty clever how the meat of the oats is separated from the chaff. See Appendix J for Oat Milling.

4. Creamery and milk processing

This plant would also produce cheddar cheese, fluid milk, cream, butter, and other dairy products. It may be more profitable to buy bulk milk than to own a dairy herd. Once again, the objective is to conserve transportation fuel by avoiding the long haul from the farm to the plant and then back to the consumer. Appendix L contains a number of topics related to milk production and processing into various products.

5. Egg Supplier

This activity implies that the community owns the egg machines, the *old biddies*. Fresh eggs sell at a premium, and the expense of long hauls can be avoided, plus some of the refrigeration costs, which use grid current from coal fired power plants for the most part. Have you ever noticed, farm fresh eggs are hard to peel once they are boiled but the older ones peel easily?

6. Bakery

I love the fresh specialty breads and the fattening pastries, much to the chagrin of my doctor. The Amana Open Hearth Bakery offered some very attractive baked goods plus soup and sandwiches. Bakery output should be sized to meet a given demand without the need for long hauls which defeat the idea of offering fresh products and saving transportation fuels. Small bakeries still exist in the nicer supermarkets and sometimes as a stand alone operation, so this is something that is indeed in demand. If a bakery could serve the needs of a county, say 20,000 people, then that would mean steady work for a few people, and yet a short haul to the most distant point in the county.

7. Fish Farm—Catfish is a favorite at restaurants that serve country style cooking. The objective here is also to conserve motor fuels and refrigeration costs.

An article on the internet by Frank A. Chapman of the University of Florida Institute of Food and Agricultural Sciences (3) gives a comprehensive view of what is involved in farming catfish. Channel cats are the primary species of farm-raised fish and are raised as far north as Arkansas, and many of the species are thought to have originated in Oklahoma around 1949, so there may be hope for other states in that latitude of the United States. Catfish are hardy enough to withstand the rigors of Oklahoma and Kansas winters but may not grow much when it is quite cold.

Catfish require a nutritional diet, as explained in the article, and a yield of 3,500 pounds per acre of water can be expected per year. The estimated time to raise catfish from the egg to food-size fish is 15 to 18 months, and this depends upon the water temperature, the density of the fish, the amount fed, and their diet, to name a few parameters. A harvest weight of 0.75 to 1.5 pounds body weight is common, and the yield of meat is about 60-65% live body weight. Please refer to the web site for more information as it is scholarly and yet interesting if you are contemplating the raising of catfish in a farm environment.

8. Cattle and Hog Feedlots

I don't mean to go overboard on this rather smelly business but there is a great potential to save motor fuel by raising the cattle and hogs locally and having them slaughtered nearby. Small specialty butcher shops usually will be happy to do the processing for you and the resultant product quality is something that can be readily controlled and observed. And feeding the home grown hay and grain will further save on motor fuel costs because the long hauls to the feedlot and slaughter houses will be eliminated.

A beef animal will dress out at about 50% of the live weight, and only 80% of that half of a critter, or 40% of the live weight, becomes USDA Choice beef. (1) In Kansas the major slaughter houses are at Emporia, Great Bend, Dodge City, Garden City (actually Holcomb, which is west of Garden city), and Liberal. Emporia is reasonably close to the major populations centers of Kansas City and its suburbs, Lawrence, Topeka, and Wichita, but the rest of the cities I have listed are a long ways from even Salina, Kansas, which is in the middle of the state. So what does this mean? A lot of animals are shipped from the farm to feed lots in the area of the slaughter houses, where they are finished or fed a ration to put the desired amount of fat and lean tissue on the animal. The finished animal is then taken to the slaughter house and processed. Trucks run 24/7 from the major packing houses, hauling boxed meat in refrigerated trailers, so here is a major expenditure of diesel fuel associated with the meat packing industry as it currently exists. If we were to run short of diesel fuel then that meat would go wanting for an outlet and the packing houses would have to curtail, maybe even shut down their operations.

A slaughter hog will dress out about 58% of live weight (2), with a 250 pound live weight hog yielding about 145 pounds of the good stuff such as sausage, chops, ham, and bacon, and oh yeah, don't forget the country backbone or ribs for barbecuing. Here again we have a long haul from the farm to the feedlot or slaughter house, and then a long haul back to the consumer of the processed meat. We can avoid much of this unnecessary waste of diesel fuel by growing our own and processing our own nearby. The Amana Colonies still have a meat shop that smokes meat and offers nice cuts of beef and pork, just to name a few products, and I think this would be a great draw to Fort Hickok, a down home meat market.

5

People Oriented Businesses

The ultimate objective of the green concept is to provide for people's physical and emotional needs in an affordable and sustainable manner. Serving people can be exhausting and frustrating, and will require workers of various ages with various skills. However, seeing smiling and satisfied customers makes it all worth while. As with any business, be it a product or a service, those who offer top quality at competitive prices will prevail. Some of the activities are for the indigent and the frail, which is part of living in a community with common aspirations and needs. We can no longer afford to dump our oldsters in a lonely rest home, and we shouldn't, so let's help them in return for their helping us when we needed it.

The list is long and most of the activities are self explanatory so I will try not to bore you with redundant information. Many of the activities relate to the tourism business, some relate to caring for those who need special care, and other businesses offer an opportunity for those with low skills and education a chance to make it on their own.

I think a lot of the entertainment activities have been inspired by my visits to Branson, Missouri, which is a great place to spend a vacation, and from a visit to the Rutlader Outpost (1), which is essentially an entertainment venue located about 40 miles south of Kansas City on US Highway 69. That plus seeing folks on tour busses stop here in Abilene for a night's rest as they traveled to Branson or back home to Denver told me we were missing a chance to share our heritage and hospitality with those travelers. The fact that they spent the night and moved on also told me that Abilene was missing a golden opportunity to be their travel destination rather than Branson because there is really nothing special in Branson that we couldn't offer in Abilene except the weather and the Ozark Mountains.

Abilene could have built an opry house like the Middle Creek Opry at Rutlader, and could have built a nice resort park where folks could golf and relax, and the distance from Denver to Abilene would be some 450 miles rather than the additional 360 miles on to Branson. Rutlader offers a show with a country artist for $18 dollars a head where a similar show in Branson would cost at least $30 or more. In Branson it is common for two older stars or artists to team up on the same show ticket, and one performs for the first hour and the other does the second hour, which is essentially what the Middle Creek Opry does when it has a visiting artist, so $18 for two hours entertainment is a real bargain.

The local hotel manager told me a few years back that tour bus trade has fallen off dramatically and that the few busses that do stop in don't have a full load. This tells me that the cost of motor fuel has taken its toll and that the people can only go back to Branson so many times without getting bored with the same old routine. The problem as I see it is too many hours in a sweaty bus seat and too few hours having fun at the destination. One more point on tour busses. People get tired of seeing the same thing year after year, which is what they are left with if they only have sights to see at a destination, but with Branson the scene changes in that new performing artists come each year. Maybe that is why I keep going back year after year, to see another favorite of mine from earlier years when they were hot. A theater at Fort Hickok could also keep them coming back if the shows changed each year, and could serve thousands of people a year who wouldn't have to drive two days to get there.

I think that Fort Hickok has something to offer, and in its own right can be very competitive cost wise. There was nothing spectacular about Pella or the Amana Colonies in comparison to Las Vegas or Branson and yet I dearly loved that trip to some genuine ethnic American heritage. We could dovetail tours of the Hickok business activities in with the commercial entertainment to provide the customer a nice experience at a fraction of the cost of going to Branson, and most of it would be the real thing rather than something make believe. Remember my first Fort Hickok in Abilene, a make believe historical theme to enhance its appeal to visitors, and yet a real live working retail development? In the Amana Colonies the history is real and it works quite well for those folks. Over a period of decades the history of Fort Hickok will also be real and can be drawn upon to please the visitors to the community.

In case you haven't heard, Pella, Iowa, a city with Dutch heritage (2), has two top quality industries, Pella Windows, and a plant that makes excellent trenchers and haying equipment, so this is probably their mainstay of income. The Amana Colonies (3) have a plant that is now owned by the Whirlpool Corp. and it makes refrigeration products. The original *radarange*, an early microwave, was designed and produced right there in one of the little Amana villages. Radar equipment uses the microwave frequency of radio wave energy, hence the name *radarange*. Fort Hickok would also have its mainstay industries, preferably green industries in the sense that those businesses work to save motor fuel by producing items locally rather than having them trucked in.

1. Child Daycare—for residents and outsiders
This is being provided because of the need of low income parents to have a day-care service while they work in the community.

2. Adult Daycare—as an affordable alternative to full time rest home care. Primary caregivers could leave a patient with the daycare center while at work and the center would provide much the same services as a full time resident in a nursing home would recreive.

3. Assisted Living—particularly for those with modest needs and modest means. The assisted living homes I know of are very expensive, possibly more so than a nursing home stay, but there is a need for a modest style of assisted living for ambulatory elderly folks.

4. Boarding House—primarily for singles who work at the community or who do not wish to do their own housekeeping and cooking. These units would be small suites in a group setting, much like in a hotel, but with some common areas for laundry, recreation, and such. People who have sold their home and are waiting to get into the next one, and those in town for a stay of a few weeks or months will welcome the chance to stay in a decent boarding house.

An alternative to the traditional boarding house would be a cooperative house where the tenants share in some of the housekeeping chores as part of their rental fee. Full times cooks would man the kitchen but coop residents could help serve meals, wash dishes, and clean the dining room.

5. Pet Grooming and Boarding Kennel

Travelers often take their pets along on trips, and they sometimes need a safe comfy place for their little ones to hang out while Mommy and Daddy go play. I get a good scolding if I don't take my Ladybug in the pickem up truck, and she is an excellent traveler. However, we leave her in a local kennel if we are taking a long trip, and it would be nice if we could take her along and then leave her somewhere as needed. Babies like to go too, especially dogs.

6. RV Park

God Help us if fuel costs don't come down. If not I may park mine this year. RVing can be a fun and inexpensive way to travel and RVs (recreational vehicles, which includes trailers) are popular with the older set. An RV park can start out small and then expand if trade volume picks up. Let's give folks an opportunity to stop in for a short stay and take in the local attractions. The name of the game is to have something that appeals to them so they will dig into their jeans for the bucks. Why should they drive on down the road when they can save fuel by stopping at the fort? And if nothing else can be furnished at the RV park, make it be good tasting water. Some places have yucky water, so we usually carry our own drinking water.

A really great RV park would have some planned activities for the tenants, at least on the weekends. The Rutlader Outpost has a nice RV park and offers shows at the Middle Creek Opry on weekends. Those who purchase tickets for the shows get a nice discount on their lot rent at the RV park, and one can walk to the theater from the park rather than having to burn fuel waiting in line in a car.

7. Micro brewery—a pub with brews, music, and food.

The nearest national brand brewery must be 150 miles away from Abilene, and beer is mostly water, so let's save fuel by making our own suds. Some of the concoctions I have tasted in microbrewery were great, some weren't.

8. Restaurant

Features include a buffet, home cooking, menu items, served family style. This is a takeoff of my Pella and Amana experience. Ethnic foods are fine but it is hard to beat good ole American roast beef, taters and gravy, fried chicken, apple pie, and hot rolls. I grew up on that fare and still can't shake the habit. Why can't Granny's cookin' be called *ethnic foods?*

9. Motel for tourists and others—simple, quiet, and moderately priced.

Somewhat of a lodge or bed and breakfast approach to lodging is what I have in mind. The idea is to give the patrons a place to co-mingle, lounge around, play pool, or do other casual things. Even a matinee dance with my disk jockey service could be thrown in. Are you ready for that! The breakfast buffets are a hit with folks in Branson, and they are reasonably priced, possibly the best deal of the day for food.

10. Picnic ground

My visits to the Sandy Lake Park (4) near Dallas, Texas gave me this idea. It is an inexpensive way to have a group gathering or a family picnic, with activities and a catering service. This park would be the backbone of the summer recreation activities at Fort Hickok, and a modest entrance fee would be charged. This is a conveient service for company parties, church groups, etc. And a lake with ducks and geese plus some paddle boats, bumper boats, etc. would be a plus.

A take off of a picnic ground would be a dude ranch that offered lodging and meals plus entertainment. Perhaps the other facilities of Fort Hickok could be used to complement a basic dude ranch establishment. A dude ranch could also cater to church groups and youth groups such as the Boy Scouts and Girl Scouts.

11. Car detailing

This activity is intended to provide those with low skills a chance to earn their own way. A pickup and delivery service might be offered or a detail job while you wait. Just today, May 16, 2007, I learned that two people can detail the inside and outside of a car in about four hours, and the cost is roughly $90 for a car and $120 for a minivan, an SUV, or a light truck. The local GM dealership is particular about its used cars and drives them 25 miles to have them slicked up, and then 25 miles back to the sales lot. So here is a couple gallons of gas per trip that can be saved by doing the work locally. By dovetailing this service in with a tour and lunch the community could offer a nice day for the customer while the car is being detailed. You have be creative and you have to hustle if you want the business, and people are hungry for quality work, reasonable prices, and a pleasant experience while they wait. Go for it. See Note 5 to view more details about what is involved with car detailing.

12. Dance club

Snacks, games, micro brewed products, and good country music, cold beer and warm(?) women, what a deal. I spent seven years chasing those beauties around the room at night, until one with a big cocker spaniel caught me. Freckles and I hit it off just fine in the recliner while Edna got ready for our date, and if Freckles hadn't liked me I would have been out of luck. In my vacation travels there have been times that I would have gladly traded a few hours on a country dance floor for all the attractions in town. Dancing can be an inexpensive and fun way to exercise.

Freckles was an English Cocker and was multicolored white with brown spots on her muzzle and forefeet. Once she was sheared the little brown spots looked like freckles, and that is how that sweetheart got her name.

13. Liquor Store

I can live without liquor myself but many travelers from the metropolitan areas are used to having their alcohol. As long as they do it in moderation and behave themselves I have no problem with selling it to them as there seems to be a good profit margin in liquor. A so-called discount liquor store can attract customers from nearby areas, especially if it can carry favorite brands of wine and liquor.

14. Gift Shop and Bookstore

Like other tourists, I need a little memento of where I have visited if the place really grabs me. I bought some overpriced jam and a few post cards plus a set of salt shakers while in Amana, and a paperback book on Amana history for 13 bucks, so it all adds up. And the profit margin is surely generous? My family and I are pushovers when it comes to gourmet food items such as cheeses, meats, and chocolate, and we buy such things as munchies to eat in the room or in the car. If Fort Hickok can sell some memories in the way of post cards, books, and treats that bear the name of the community I think this can be a part of the tourist trade, even for tour busses as those folks need small items to tuck in their luggage or to carry on the bus.

15. General Store—for locals and for outsiders

Groceries and other basic needs such as hardware, lawn supplies, and repair parts and supplies would be offered. This would also act as a convenience store for those who need small items such as camera batteries, film, snacks, etc. I can remember driving 25 miles one evening to buy a nasal inhalant because I was plugged up and couldn't breathe though my nose. A local store that was open would have been a welcome sight. Farm cooperatives sometimes have a grocery and general merchandise store, and members of the coop enjoy the savings of the low overhead.

Stores that carry the hard to find tools for the kitchen and the garden are a hit. My wife bought a small wedge shaped hoe and paid dearly for it in a hardware store, but she has enjoyed years of using that little hoe, which is much easier to manipulate than the standard big ones with flat blades. The local hardware store carrys the Radio Flyer line of wagons, scooters, and sleds for the kids, and those are hot sellers at Christmas time. Having something special in your store gives people a special reason to come see you.

16. Country Opry

Shows are usually offered on Saturday nights, and hopefully some nationally popular artists might stop in for a gig or two. The Rutlader Outpost has the Middle Creek Opry (1), which is a country music show staged in a simple metal building. It has 7,000 square feet of floor space and can seat 500 patrons on a level concrete slab. The chairs can be quickly cleared away to allow other things to happen in the space, which can be rented by the day for other reasons. I find this to be a more versatile structure than the formal theaters in Branson, but it is not quite as comfy and is the least bit showy. Still, it is a nice way to get started in a venture because the building can also be used for other reasons. I think a capacity of 1,000 patrons would come nearer to being profitable than a capacity of 500 because the popular visiting artists will want a minimum amount to stop in and put on a show.

Isn't it less fuelish for a star and the small troupe to travel, say 500 miles, to perform for hundreds of people, than to have those hundreds of people drive that far and more in some cases to see the show? The nice thing about Branson, Mo. is its location, somewhat centrally located and within a day's drive of some major population centers. Contrast this to going to Las Vegas, which usually dictates a plane ride. A big 747 gulps fuel at the rate of one gallon per 750 feet of travel, I once heard. Wow, and I though my old truck was thirsty. Fort Hickok can help tourists to avoid this wasteful expenditure of fuel by offering them entertainment and other recreational services close to home.

17. Catering Service

This activity would be used to support the picnic grounds and to serve other local needs such as special gatherings. Top quality catered food is a rarity so if someone wants to deliver a nice meal for a competitive price I think there will be plenty of takers. This is a good family oriented business because mom and dad can do the work and the kids can get the profits, or eat into them on the way to serve!

18. Healthcare Services

Minor emergency care and eventually a small hospital. Why drive 25 miles or more to see a doctor or a dentist? If a VA nurse or doctor can visit the community on a regular basis this is a plus and saves fuel. In other words, bring the doctor to the people rather than all those people driving 80 miles one way, as I do, to visit a VA center.

19. Dude Ranch

This is a fun way to spend a few days lounging around, riding horses, swimming, and taking part in the various activities offered by the ranch. These venues are usually for active adults and children who have energy to burn, but a program for older folks could surely be designed and offered. Dude ranches provide room and board plus recreation opportunities. I think this activity could be included as part of the resort lodging that would be at the Fort. The synergy of making things work together to present a vacation package at a reasonable price adds leverage to any one part of the overall package, and makes more jobs for local folks who don't want to drive into the far cities for work.

6

Miscellaneous Businesses

These activities are intended to conserve motor fuel, to provide an alternative to non-renewable fuels for the home, and to provide recreation in the form of market places. Fruits and vegetables are perishable and require refrigerated transport in the hot months, so that is an additional fuel cost that can be saved by growing and marketing them locally. Refrigerated trucks typically use an engine driven compressor for the cooling, and the same goes for refrigerated train cars. These beggars also like fuel as well as the engine under the hood that takes the rig down the road, so here is additional consumption of motor fuels that can be avoided.

1. Fruit orchards, berry shrubs and bushes

The output can be processed or sold fresh to customers. Apples grow well in Kansas, as do wild plums, currants, gooseberries, and raspberries. Have you ever tasted bumble berry pie? It is a crust pie with a variety of tangy fruits for the filling. That with a scoop of ice cream and I'll fight you, for my piece that is.

2. Community vegetable gardens—for use by residents and outsiders.

The heavy tillage work would be done by machines for all the plots and the watering and detail work would be done by the customers. Small community garden plots seemed to be the go in Frankfurt, Germany when I was there in the service (1967) and even in the colder months one could find people puttering around in the dirt. My wife calls it dirt therapy, and it works.

Small gardens provide a way to produce the specialty crops that do not warrant a full scale commercial operation and also provide people a chance to have their own small business. For example, horseradish, rhubarb, and gooseberries are nice offerings from small gardens.

3. Farmer's market—a chance to sell local produce and to socialize with folks.

Homemade items such as pies, cookies, jams, and sauces are also in good demand at these markets. This is a good draw to the area if a large city is nearby, and the produce from the small community gardens can sold this way if desired.

4. Flea Market—with emphasis on recreation, entertainment, snack bars, rest areas, and other people amenities. Heck, just rummaging thru all those goodies is recreation.

5. Cordwood sales and fuel pellet sales—especially if the community has its own wood lots and can make its own fuel pellets. It appears that even switch grass and municipal solid waste can be made into fuel pellets. Pellet form permit's the fuel to be bagged, stored, and easily handled by an individual.

I once read somewhere on the internet where people in Europe were using biomass to fire boilers, and the hot water was being pumped to a small community for domestic space heating. Perhaps this could be possible at Fort Hickok for some of the group home settings.

The Amana people grew nearly all of their fruits and vegetables, and apparently there were gardens in place of lawns with grass. This from a resident of one of the villages. The houses had trellises attached to the exterior walls for holding the grape vines, and around the house would be a narrow band of soil to be used for cabbages and such. The grape vines helped to temper the air with moisture and to cool it as a breeze passed through.

7

Alternate Energy Devices

There are ways to do things without using the traditional forms of energy such as grid power, more often than not generated by a coal burning power plant, or natural gas, which is essentially a methane gas, and of course to make do without petroleum based motor fuels. An added virtue of using alternate power that is produced locally is that it offers a backup if the grid power or natural gas supply is interrupted by storms, flooding or acts of violence. In this age of terrorist attacks it would be comforting to know that the power is still on even though the grid took a hit miles away. I also propose these few items as a way to reduce our carbon footprint.

1. Windmills for pumping water—New mills are still available and a mill can provide a large quantity of water as long as the wind blows and the well doesn't go dry. This water would be used for livestock, for irrigating lawns and gardens, and for other applications where raw untreated water is safe to use. In some situations it may be feasible to treat and filter the well water so it can be potable for direct human consumption. If the mill could be located on high ground, which it should be to catch better wind speeds, the water could be saved in a reservoir tank and piped down the slope for use. If the pressure is too low then electric booster pumps could be used as needed.

2. Wind generators for electrical power—primarily those sized for domestic use, but the big boys being on the property would be an extra source of income. I was really surprised to learn of the power the utility sized machines can deliver, on the order of 1.86 megawatts. In terms of shaft horsepower that would be a whopping 2480 HP, which would equivalent to a large 16 cylinder diesel generator unit. Can you imagine that much power on a stick 100 feet in the air?

If you have been on I-40 west of Oklahoma City, Oklahoma you may have noticed the wind farms. A utility company will pay the land owner to lease the land for the right to place wind generators on the property, and if the generators are installed the utility will pay the land owner an additional fee, which is quite substantial per machine. Kansas is also a windy place, as you may have sensed from reading my long winded article, and other states also have great potential for wind generated grid current. If the units are small in size, the community could use them as needed for commercial activities such as powering a greenhouse or providing lights for a small factory or store.

I use the fluorescent light bulbs here in the house and a bulb with 60 watts of output only consumes 13 watts of input power. It is not quite enough for me to read by, but it is fine for general illumination around the house, and a wind driven generator could power a bunch of these curly fried little lamps in a residential structure, especially in the case of rental units where the tenants do not pay the utility bill and are careless about shutting off the lights when they are not needed.

There are two basic types of small wind machines, those which deliver power directly to the power grid, and those which charge storage batteries. Both types use some processing circuitry to get the correct frequency and amplitude of the output power. An individual or small business can also use a DC (direct current) system, off grid that is, and this involves a string of storage batteries. For example, my RV trailer uses a 12 volt DC battery operated system for the lighting, the radio, and the water pump, so 12V can certainly power much of a home. Other DC voltages range up to 48 volts or more, which is a rather powerful force. However, I envision using 110/240/480 volts AC (alternating current) for Fort Hickok so the electrical devices can be compatible with grid current, but other voltages could be used if it were advantageous to do so.

3. Hydrogen production—Hydrogen gas can be fed to fuel cells to generate electrical power, so it is a way to store energy for future use. It can also be used much like LP gas or natural gas, and can be used as a motor fuel. What I have in mind for hydrogen generation at the community is the electrolysis process, which requires electricity, and if some small scale water power projects were available that would be a green way to make some cheap electrical power. However, a wind powered generator will do nicely.

Hydrogen gas is not as convenient to store as methane or Liquefied Petroleum (LP) are but it can be done, particularly for a stationary application. The advantage about using water power to make hydrogen gas is that the potential energy can simply be stored as water behind the dam and used as needed to generate electrical power, which is the driving force for electrolysis. With wind power one must either make hydrogen only when the wind blows or by using electrical energy stored in batteries when it doesn't.

At the moment the big problem seems to be finding a safe and cost effective way to carry enough hydrogen in a vehicle to give it a practical range. If a mass transit vehicle operated off an electric tether that was powered by a stationary hydrogen powered device, be it a generator or a fuel cell, then the problem of mobile storage could be avoided, but that project is beyond the scope of this work. Such a device might resemble the old fashioned electric trolley with power lines overhead. Just wanted share with you another way to do the same thing with less use of hydrocarbon fuels, another way to skin the cat. I firmly believe that such thinking and innovation is essential to a sound and affordable national energy system.

4. Methane production—This process uses a digester to convert plant waste, municipal waste, and animal waste to a gas that contains a lot of methane, which is essentially natural gas like that bought from a utility. It can be done on a small scale, right at the home, or on a community or commercial scale. The process is called anaerobic digestion and has been done on a small scale by the do-it-yourselfers for decades now.

At present my town lets folks take their grass clippings, plant wastes, and leaves to a compost pile. It is a way to take the load off of the landfill, but the piles of rotting material emit a number of gases to the atmosphere. I think that using the digesters can avoid that problem and could produce some usable gas.

5. Wood Stoves—cordwood stoves—to be used for heating homes, offices, and small structures. This is a dependable backup feature although it is not as convenient to use as a pellet stove. The community woodlot could supply a portion of the firewood but some may have to be taken from the naturally occurring trees in the area by selectively pruning in congested growth areas. An alternative to burning cordwood is to burn corn kernels, which is also possible to do in some pellet stoves. In a common wood stove the corn is placed in a small canister with holes

in the sides to permit good air circulation, and the corn burns with a low blue flame.

6. Pellet stoves—Various types of pellets can be used and it is important to use a fuel that is low in ash and tars so they won't collect in the stack and start a fire. My stove seems to be very efficient because after burning 40 pounds of oak wood pellets less than a cupful of residue remains in the stove when we clean it out each morning. And some of that cupful is unused pellets that missed the firebox when they fell from the storage bin.

Top quality pellet stoves can be expensive to buy and to install, and most of the better ones depend upon electrical power to operate cleanly and safely. Mine has a circulation blower to get the heated air out into the room, a combustion blower to force a draft of air to the burner, and an electric auger that measures the pellets into the firebox at a controlled rate. Perhaps the greatest improvement for pellet stoves would be to design one that could also tolerate pellets made from grasses and other materials that contain more tars and ash. One would also have to have a chimney system that would tolerate the buildup and could be easily cleaned out. There is, by the way, a pellet stove being offered that generates its own electrical power. My pellet stove has an automatic thermostat control, which helps to save energy when the room temperature reaches the setting.

Someone at Fort Hickok needs to invent, patent, and produce a moderately priced pellet stove, particularly one that could tolerate various types of feedstock. So here is a chance for a handyman with some inventive genius to make a contribution to the greening of America by inventing one of these stoves.

8

Affordable Housing

Affordable housing is right sized and thrifty to rent, or own, and operate. This topic deals with designing a home to be energy efficient and providing an owner-ship or leasing arrangement that will minimize taxes on the property and aid the passage of one's equity in the home to heirs. Designing houses has been a pet project of mine for years now but I would like to defer the details to Appendix C. You'd be surprised what a little noodling up front can do for the homeowner and an objective of the Fort Hickok community is to give a better bang for the buck when it comes to housing. What, you mean you have never noodled?

There would be various types of housing at Fort Hickok, single family dwell-ings, duplexes, fourplexes, and group homes, as with a military or college setting, including a dining facility. To the extent possible the units will be built to share adjacent walls in order to decrease the cost of heating and cooling them.

For years it has been advocated by those who stand to gain from constantly *trading* up in housing to buy a *starter home*, live in it a while, and then use the equity in that home to move up to a larger and more expensive home. Yep, I have done that very thing, but there comes a time when one needs an *ender home*, one that is affordable in the senior years and not a burdensome status symbol. I think that Fort Hickok can accommodate that need by offering the various types of housing, which is explained in Appendix C.

Realtors, builders, and others have also preached that real estate is a good investment, and yes, in most cases, it can be a very profitable investment. This comes about because with a down payment that is only a small portion of the total purchase cost of the unit one can control a much larger investment that increases, not by the equity appreciating, but by the market value of the home increasing. This is known as leverage, and that is great, unless the market turns against you and you have lost your means to pay for the house. Then that nice investment can turn a bit sour. Folks at Fort Hickok shouldn't have to play that game and shouldn't have to pay a realtor to spend their hard earned cash on a new home. Those things can be avoided by buying and selling through the cooperative or company that controls housing in the community.

9

Healthcare Services

Hopefully retired or semi-retired nurses and doctors could provide some rudimentary health care as needed to the residents and to visitors on a low cost basis. A community financed doctor's office could also be provided so that the nurses and doctors could see patients for a reduced rate. If the demand would sustain it, a small hospital could be built, as a town of less than 2,000 often has a small hospital, a few doctors in town, and a dentist or two. Oh yeah, and don't forget a vet for the babies.

Something additional to be considered. The jobs offered by large corporations are attractive to many over those offered by small companies because the big companies can offer group health insurance but the little ones just cannot do that and stay competitive. Year by year we are moving closer to a national healthcare system, or socialized medicine as it is sometimes called. If this were the case then there would be no need for the employer to offer insurance as a benefit to attract employees, and folks could stay in town and work close to the house. So payback time may be drawing near for the American worker and it is sweet justice after seeing our jobs go South or across the pond to the third world countries with *good cheap help*, as a work mate of mine used to call it. Bless his heart, he was a union man from Ohio, and an excellent worker. And he had a point thar judge!

As much as many Americans dislike the idea of socialized medicine, and as much as the healthcare providers fight it, we still have that very type of system to a great extent with certain groups of people. My understanding of *socialized medicine* is one where doctors and practitioners work for the government, as in Canada and in England. This arrangement has its downside as well as its advantages. Americans take pride in being able to choose their healthcare providers and yet complain about there being no competition, which might bring down the prices. Yep, there is little competition, few alternatives to standard health care offerings.

Medicare is a modified form of socialized medicine or nationalized healthcare, with the practitioners being private business people and the patient having some choice in which one to use. However the Medicare officials have a lot of say in what is done, how it is done, and how much it will cost. Those in the armed services have access to what is essentially socialized medicine in that the practitioners work for the government, and the facilities and prescription medicine is owned by the federal government. As a veteran I use the Veterans Administration (VA) healthcare facilities in addition to having my own doctor in the private sector, so I have the best of both worlds. Indigent people, especially children, have access to Medicaid, which is much like Medicare except that the cost is borne by the taxpayer to a large extent.

Why couldn't Fort Hickok have sort of a privately owned Medicare system, with the patients paying some of the costs of their healthcare and the system paying some of the rest. This sounds like group health insurance, and to a certain extent it is, but the carrier would not exist to gouge the patron or to make big bucks off of Susy having tonsillitis.

Commercial group insurance carriers and even Medicare seek to limit the amount paid for a claim (a service of some sort, or a prescription drug, just to name two types of claims) and they mean well. Supposedly the patient, or the payer I should say, benefits from this attempt to hold down the cost of health care insurance, or the premiums, but from what I have seen of doctor fees the claims are marked up knowing full well that the carriers will only pay a part of them, so who is kidding who?

Does limiting payments really ease the true of cost of providing the service or products, as seen by the health care industry? I think not, because the cost of educating the doctors and other staff, the cost of the diagnostic equipment, and even the cost of the plant are not subject to any limitations imposed by commercial insurance carriers or the federal government. My point is this, the true way to hold down the cost of healthcare is to address the factors that make it cost what it does, and I have listed a few of those costs.

If Fort Hickok could somehow provide monetary assistance to the nurses, the doctors, the technicians, and other skilled staff members, and if in return these people agreed to work for a lesser wage for a period of time, then that approach would tend to hold down the cost of healthcare in the community. The government assists new doctors by having them serve a period of time in public service, such as in the armed services medical system or the VA healthcare system, so I think the approach is well grounded in experience. Perhaps some of the profits from other business activities in the community could be set aside as financial assistance to train these healthcare workers. Additionally, subsidized housing and discounts by participating merchants in the community would also serve to sweeten the pot for those participating in the program.

Relative to the cost of a college education, a prominent family in my childhood home town had a bunch of kids, and all of them went to college if I remember correctly. The parents had a bank and could afford to educate their kids, one of which became a doctor, others became lawyers, and some majored in business disciplines. This education went on for years as the kids were stair stepped in ages, so the parents bought a house at the University of Kansas, at Lawrence, and the kids would live in that thing while going to school. The wives got PHT degrees (Put Hubby Through) so overall it was a smart move to buy the house. Perhaps Fort Hickok could also buy or build houses or cooperative residences at one or more major schools for their students to live in. This in effect would be helping those kids get their education.

There was a time when church groups ran the hospitals, and they were there to heal the patient and not to get fat in the process. Contrast that to what we see with corporately owned facilities. Yes, modern equipment is costly, and that may have caused some of the non-profit hospitals to privatize and incorporate so they could raise more capital to equip the facilities. However, there are still some top quality non-profit healthcare facilities and I think that Fort Hickok should aim to start this way, with a non-profit healthcare system.

Adult daycare is apparently a way to reduce the cost to those who might otherwise be confined to a rest home on a full time basis. The participant is taken to the daycare center for a period time while the primary caregiver goes to work or does other chores, and while at the center the person receives the usual assistance that is provided to a full time resident such as baths, meals, therapy, and entertainment. This strikes me as a nice service for the participant and for the primary caregiver.

10

Communal Living Concepts

Fort Hickok would have a limited communal living concept, somewhat of an eco-village but at the practical and feasible level. The communal living style of early Amana was essentially a communist form of economy where a corporation owned most everything such as the factories, the housing, and even the doctor's cars. However, these wonderful people were pacifist in nature and simply wanted to be left alone to pursue their religious beliefs of simplicity, self-sufficiency, and fundamentalism. They did not seek to usurp the law of the land although they did have some rules of their own that community members were expected to abide by.

I think a modified form of communal living, called an intentional community, socially speaking, would be more appropriate and would appeal to a wider audience than a strict religious sect would. The preferred attitude of a community member would include a desire to be self-sufficient and conservative, particularly in respect to basic forms of alternate energy, to the extent that is feasible for the locale and situation, and a willingness to cooperate with others to make things better for all concerned. If friends can do little things for each other there is no need to pay to have them done. Keep in mind that the primary reason for creating the Fort Hickok community is economic in nature and the ultimate objective is to utilize alternate forms of energy and to reduce the amount of fuels needed for transportation in order to better the lives of the residents. Hopefully the development would serve as a role model for other communities to emulate to the degree that suits their needs and situation, and in a subtle way it is very much a green community.

For a look at what I hope is not Fort Hickok someday see web site http://www.thefarm.org/lifestyle/cmnl.html. (1). *The Farm*, written by a former member of the commune being described, was somewhat of a social experiment or a rejection of materialism and the everyday life of America in the 60s and 70s. I didn't see this spirit of futility when I visited the Amana Colonies although they had a major revision or *great change* in the early 30s after realizing that the old concepts just couldn't serve the needs of their community like it did in the beginning, mainly because the modern residents had seen too much of the outside world and wanted some of it for themselves, the material things, and the religious and social freedom. Socially and economically speaking, Fort Hickok would be much like any other small town except for the common desire to save motor fuels and to utilize alternate energy sources.

A residential city is essentially a bedroom city or a commuter town, and I don't see this for Fort Hickok as it should be somewhat self-sufficient in industry and commerce. The Wikipedia encyclopedia has a brief article on *Residential community* www.wikipedia.org/wiki/Residential community (2). Surely more can come of being self-sufficient and green than a city of commuters. Some of the residents may have to work in a larger town, as is the case in the Amana Colonies, but the objective is to provide local employment for most of those who wish to work close to home. I think the need of healthcare insurance will continue to be the primary reason why people seek jobs with major companies that can offer healthcare insurance as a benefit. However, as I have already mentioned, there may be ways to moderate the cost of basic healthcare provided right in the community rather than having to drive an hour or more to be seen by a doctor or a nurse.

What is an ecovillage? The new Fort Hickok will probably be called everything from "free love ville" to "tree hugger heaven", but the intent is to simply conserve fuel and get off the petroleum oil barrel by using other forms of energy that can be produced locally. Once again, Wikipedia has a brief article on ecovillage (3), and such a village is intended to be a socially, economically, and ecologically sustainable community. Sounds like we are getting closer to what I envision for Fort Hickok, the green community. Small rural towns are essentially self sufficient in the social sense, from my experience, even though I am disappointed at the lack of night life or commercial entertainment in small towns. We can alleviate that by purposely designing entertainment opportunities into the norm of the community, and I have listed a number of people oriented activities in Chapter 5 People Oriented Businesses. In other words we are going to purposely design and build boredom out of the way at Fort Hickok.

If the new community is to be an ecovillage, then let it be a modified version of those who reject amenities such as city water and sewer, and utility grid power. We can make much of our own power and heating fuels locally, given time and effort. Take a look at Chapter 7 Alternative Energy Devices and Appendix D Methods For Making Biofuels, and you will see that there are a number of ways we can avoid being held hostage to big oil and to insensitive utility companies.

Perhaps you aren't old enough to have lived in a small farm town when the power was produced in someone's garage by a decrepit generator and the telephone system was a ring up one with a *central* and used big dry cell batteries. Natural gas was available if the little town was near a major pipeline, but otherwise it used bottled gas and wood for heating. Hey, I grew up in a little town, about 20 years after the local generator in grandpa's garage broke down for the last time, and we didn't even have central heat or air because the old houses weren't designed and built with such amenities. And we made it, fat, dumb and happy. There is no need to live like that in Fort Hickok just because we are stingy with motor fuels or use pellet stoves, so never fear, it would be a modern and affordable lifestyle, and it would be a sustainable, environmentally friendly one.

As the ecovillage article points out, these villages rely on green businesses, clustered housing, and renewable energy to minimize their ecological footprint, and I think this can be done to a point with Fort Hickok. The autonomous housing, or that which does not depend upon a utility for power, water, and such, may not be feasible, but as I have already mentioned, locally provided utility services can be designed into the little town. And provisions should be made for bike trails, walk paths, and a shuttle during certain hours, all which would serve to reduce the amount of fuel needed to jump in the car and go to get a burger, a loaf of bread, or a six pack of hooch. Shouldn't drink and drive anyway as it can be bad for your health and can stunt your lifespan!

An intentional style of community might be appropriate for Fort Hickok, and with this type of community organization the members share a common goal, such as conserving motor fuels and domestic energy. Some communities interview prospective new members and give them a probationary period to determine if they are right for the community objectives and outlook. One way to do this at Fort Hickok is to encourage new residents, for that matter, all residents, to invest in the cooperatives and other business concerns of the city and to support them with their capital, patronage and labor. It is not for everybody, but for those who wish to give it a try I think there is much to gain from it.

The Wikipedia article on *Intentional community* (4) lists a number of housing arrangements such as cohousing, land trusts, and housing cooperatives. Let's look at a few to stimulate the gray matter into finding more affordable ways to enjoy housing.

Cohousing (5) involves private homes that have a common dining facility, and the members of the community take turns cooking the meals. I see this being done in a modified form at Fort Hickok with the cooks and dining room attendants doing it full time as employees of the community organization that owns and manages the cohousing facilities. Some of the residents would live this way and others would live more independently. Different strokes for different folks, and needs.

Ownership can vary from privately owned residences, to condominiums to housing cooperatives, so there would be flexibility and choice in the occupancy of housing in the community. The housing cooperative may offer some tax breaks and advantages that individual home ownership cannot offer. The *Cooperative (6)* article in Wikipedia gives a simple description of what a housing cooperative can consist of, and the residents own the housing collectively, much as with owning stock in a corporation that might own the housing. Given this, the residents have a say on the management and maintenance of the housing, and when it is sold or passed on to another tenant or party there is no need of a realtor and no need to rip off the next tenant. In the United States most housing cooperatives are organized as Limited Liability Partnerships (LLPs), but other legal forms of ownership can be employed if advantageous to do so.

In closing, some people may do better by simply renting and not having the headaches that come with home ownership, and this is particularly true with single people, young couples just starting out on a limited budget, and older people who don't want to stand the cost of repairs, lawn care, and the other responsibility of home ownership. I think my design has taken such needs into account and more on this can be seen in Appendix C Housing That Is Right For Your Needs and Tastes.

APPENDIX A

Fort Hickok, Where Are You?

Furthermore, what are you? More in a minute on that point. An optimum location for the community would be near an interstate route or near a well traveled state highway. Abilene, Kansas is located in such a manner, being alongside I-70 and having Kansas Highway 15 running right through it. Trucks and travelers pass thru this area, and some of them stop in for rest and food or to see the local attractions. I think the design concept for a green community could be applied in a number of states across the USA, and the climate and soil of a given locale may dictate what types of agricultural activities are feasible to pursue.

The Amana Colonies in Iowa lie seven miles north of I-80 and consist of small villages. In the early years the total population of the colonies was about 1,250 people. Currently there are also Amana shops along the interstate exit to the old colonies that offer products from the Amana factories and stores, and this development is referred to as Little Amana, which may account for the eighth village. The Amana pioneers prudently bought the land of what is now Homestead and set up the village there, which is a bit south of the biggest village, simply called Amana. Owning Homestead let them access the railroad that ran through the area to ship their products out and to bring in customers and tourists.

An additional advantage of being close to good roads is being able to readily ship the bulky products such as the grains out of the area. Access to a railroad may also be advantageous, depending on the business makeup of a given community. A nearby river or active creek would also be a plus, as would deposits of clay and stone. Rolling terrain and fertile well drained soil are also very desirable traits for the development, and some high ground on suitable pastureland for the livestock would be great for the cows and the wind generators. A farm windmill will do some pumping in a valley but would pump faster on a hilltop.

A community of 1,500 people would be a good size to start with, and if the folks were scattered across a cluster of small villages this can be an advantage, and to a certain extent a disadvantage, depending on what one is looking for in a community. An upside of having several small villages is that there would be little need for the overhead associated with large numbers of city employees, and the sewer system for each town could be a simple system of lagoons. See Appendix F for information on small waste treatment plants. A downside of a small town is that there are few amenities to offer like in a larger town because if the town is too small a grocery store or a decent restaurant probably cannot survive, and the streets can get pretty rugged without a budget and staff to keep them in decent order.

Having several little towns and villages permits some modularity, which would help to promote the concept of conservation. One could have a village that related to the farming activity, a bunch of farmers, and another village could be associated with the industrial activities, that is, the manufacturing and processing plants, another specializing in retail and recreational activities, and for those wanting some peace and quite, one or more quiet villages that are basically bedroom communities. The latter type of development would be mainly for retirees or would be a rustic resort community for the more affluent class that can afford such luxuries. Places like Branson, Missouri and Las Vegas, Nevada have some very nice resort developments, at some very nice prices. What I had in mind would be a simple and inexpensive place to come and play and rest for a short time, perhaps on a time sharing basis or on a rental basis.

The business organization of Fort Hickok would essentially be one or more diversified corporations, an assortment of businesses in the legal and economic sense, and one with a sizeable amount of housing for residents and hostelry for visitors. Included in this mix of businesses and housing areas would be a number of farm tracts and industrial activities, which comprise the core or backbone of the self-sufficiency and conservation concept. The Amana Colonies were purposely designed and constructed for the same reasons. They even went so far as to build a mill race (7 miles long) to bring water from the Iowa river to power their mills, and this little canal still exists today after being rescued from the ravages of a major flood in the area. I am not aware that it is being used to generate power but it wouldn't take much to do so if the need arose.

Many of the early farm villages in Kansas were started along rivers because the water was needed for livestock and domestic use, to feed the steam engines on the railroad, and to make power for grist mills. This rudimentary form of power was simply a water powered mill shaft that drove the grinding processes. Relics of the old dam sites still exist in places but for some reason there is no interest in rejuvenating this small scale water power. I think folks are missing a good opportunity to have some clean energy to drive electric generators, but maybe someday they will again see the virtue of it.

I think it best that the city be built from scratch because trying to convert an existing one to use the features intended for Fort Hickok would meet resistance to change and would be impractical because many of the structures aren't shaped to provide optimum heating and cooling efficiency. For example, a "C" shaped house has relatively narrow rooms with a lot of exterior wall space that represents a gain of energy during the cooling season and a loss during the heating season. The most efficient shape for a home or structure in terms of wall area per square foot of floor space is a square. Square homes are not very attractive in comparison to other shapes but there are things that can be done to improve their exterior appearance.

How much land area would be required for a city of 6,000 people? Yesterday I went driving in the rain and measured the dimensions of the main body of the city of Abilene. It turned out to be roughly a square tract of land, running two miles from east to west and 2 miles from north to south, so that is four square miles, or four sections of land to a farmer, a section being a mile wide by a mile long and containing 640 acres.

Four square miles of land comfortably houses the residents of Abilene, and has ample room for industrial tracts, includes two railroads with generous right of ways, and space for a county fairground, the home of the Wild Bill Hickok Rodeo every August. There is also room for churches, schools, parks, shopping, and the usual things you might see in a city of 6,000.

It is not necessary that the plan for Fort Hickok be square or rectangular in shape, but a square layout will give good efficiency in terms of the length of roads, sewers, and utility lines. Housing need not be concentrated in one bunch but, having most of the houses in proximity to each other also helps to improve the revenue per mile of the utilities. I think it is important that the layout be designed to facilitate the use of basic mass transit methods and so that work or shopping is conveniently accessible to the residents.

Abilene contains a variety of city block sizes, with the earlier and more modest developments being in the south part of town and the later and more affluent neighborhoods being on high ground in the northern part of the city. The dimensions of city blocks vary some, but a layout of 300 feet by 300 feet or 300 feet by 400 feet is popular, and the size must have depended on the developer. Streets are typically 50 feet wide and alleys are 15 feet wide, that is, if there are alleys. The later developments have no sidewalks in front of the houses and have no alleys, which means that the homes have front entry garages and homeowners put their garbage dumpsters on the curb for pickup. My proposal for the layout of Fort Hickok seeks to minimize the space needed for alleys and tries to minimize the role of the private auto. Therefore, a minimal amount of covered parking in the residential areas will be provided but only open air parking for visitors and extra vehicles will be provided.

Some of the housing will feature residences around the perimeter of a rectangular block with a center area for walk and bicycle paths plus gardens, tennis courts, and other recreational provisions such as picnic shelters and lawn games such as shuffleboard, horseshoes, and croquet. Exercise equipment will be provided for those who are inclined to use it, and will be stationed along a walk or jogging path or in a club room. These kinds of amenities may tend to increase the land area needed for people but the benefits justify it.

The lot sizes and block sizes in Fort Hickok will also be varied depending on the type of housing in a block. One trait I like about the little fort I designed for Old Abilene Town was that entry to the buildings was from the interior and the outside walls were effectively barriers to the street, which was also the case with the old military forts like Fort Larned in western Kansas. Some of those buildings had gun ports in the heavy stone walls and very few had any windows on the outside as a precautionary measure against Indian attacks, which seldom came. Such an approach might be good for an open mall shopping center or a housing development for families with small children because the limited access and exterior walls would provide a measure of safety from being hit by automobiles.

I have a plan in mind that would have the city straddle two crossroads, and one would be an interstate artery and the other a well traveled state highway so the little town would get good exposure to people passing through the area. If a community of 1,500 were developed on each corner of the crossroads then the development could be modular and could be built piecemeal over a period of years, which would likely be the normal pattern of development. Let's call the four developments quadrants as each is a fourth of the total plan. One quadrant could specialize in retail, another in farming activities, another in industry, and another in a resort and travel type of industry. Each quadrant would have its own basic shopping with stores located in strip malls that would have exposure to traffic coming thru the area, and would have its own residential areas. A master plan for development that would be backed up by zoning regulations would ensure that the city grew in an orderly and desired fashion.

A strip mall can be of the courtyard type with parking in the middle of the development or can be a simple string of stores facing a long parking lot. Either way works but the type that encircles the parking area provides more opportunity to define areas for gatherings, which would complement the tourist atmosphere that is intended for some of the shopping centers.

Common services such as a hospital or community parks and recreation sites would likely be concentrated in one of the quadrants, because in a small town of 6,000 it is no problem to get to the hospital or to the ball fields in a few minutes by car. The residential areas will also have some small tracts for exercise, recreation, and just plain loafing around in the leisure hours, so there would be little need to travel across town for these everyday activities.

The manufacturing industries could be located on the outskirts of the town, with the smelly ones being down wind, which would be to the north of the town for most of the continental United States as the prevailing winds are usually from the southwest. A tract of three acres would be sufficient for a milk processing plant or a small grain mill, but some activities may require a larger tract because of the space needed for storing the input feedstock or the output product. Some of the alternate energy applications may need to be spaced widely for safety reasons, as is the case with an oil refinery. Who wants to snuggle up to a refinery and lose his shirt because the refinery caught fire or exploded? Hopefully this plan will avoid the need for more oil refineries, so I merely used that as an example of a volatile and potentially dangerous type of plant. And they are stinkin' things anyway!

Farming activities will require fields for growing the crops and space for keeping the animals involved. If done right these farms can be used as stops on tours of the area, as could some of the factories. Perhaps the least hospitable sites would be the energy plants that use high temperatures and pressures to convert biomass to fuel products, but I don't see there being anything all that perilous to visiting a flour mill, a milk processing plant, or an alfalfa mill as long as the site makes provisions for tours by visitors. Companies have tended to limit access to their plants by visitors because of the liability associated with it but I think a well designed plant could be used as a showpiece for demonstrating the new concepts and the value of making the needed products locally rather than having them shipped in from distant points and consuming fuel that is needed for other reasons.

Industrial applications usually require some healthy doses of power, and many of the big electric motors used call for 480 volt three phase power, which may be beyond the capability of locally generated power. If intense or concentrated heat is needed then natural gas may be the fuel of choice if it is available, but the alternate energy processes I will speak of can also develop a lot of potent fuel for such applications. And if some of that fuel could be used to drive electrical power generators then the high voltage power can also be made locally. Manufacturers such as Caterpillar and Ingersoll-Rand make engine driven generator units that are quite large, and the engines can be designed to run on diesel, oops *biodiesel*, natural gas, or possibly a synthetic fuel, which is a product of a process such as pyrolysis that is used to convert biomass to energy products. These big babies can be found on the internet by looking up the company name and then going after their products. (1) (2)

APPENDIX B

Production? How much is enough?

A primary objective of Fort Hickok is to produce things to be consumed by residents of the community or by nearby customers in order to avoid unnecessary use of motor fuels for transport. As a minimum, the various activities should strive to produce an amount that satisfies the local demand, and as a maximum only that which can be sold within a reasonable mile radius should be produced. A key premise of this idea is that the feedstocks would also be produced locally or nearby, thereby avoiding the long hauls of products from the farm to the factory, and then back to the local area for consumption.

Let's assume that the number of residents is 1,500 to start with, so the per capita consumption per year for a given product multiplied by the number of consumers should give us a good starting figure. Keep in mind that some people may purchase their goods elsewhere, but for planning purposes this seems to be a practical way to determine the minimum capacity needed for a given product. Perhaps an initial capacity for 6,000 people would generate some surplus to sell for cash during the startup years, but it might take some doing to grow that large, and a larger town would involve some overhead that a small community doesn't have, so the economic efficiency of the town could suffer.

As to the maximum miles a product should be shipped outside the community, that again depends upon the per capita consumption and the population density of the area to be served. For example, in western Kansas where the population is sparse, one might not want to run all over half of that end of the state to make a sales quota, but in the northeast and southeast quadrants of the state, which are much more populous that the two western quadrants, one might have only a short haul to sell a given product. And if the community were close to a major population center, such as Salina, Hutchinson, Wichita, Topeka, Manhattan, Fort Riley, Lawrence, or Kansas City, Kansas, then the maximum output might be considerably higher than if the production was located out in the boonies. Oh yeah, the coyotes out in the boonies love tomatoes and melons, but the rascals don't pay up.

So to sum it up, the minimum annual production would be enough to meet the local demand and the maximum would depend upon the population density or demand in the surrounding area. Let's assume the maximum distance one way for a truck delivering a Fort Hickok product would be about 100 miles. Products that do not need refrigerated transport to keep them fresh, such as cornmeal or flour, can be shipped further than a perishable product such as milk, cheese, and fresh fruit and veggies, but one must guard against the temptation to feed the world just to make that extra buck because driving long hauls simply defeats the purpose and intent of the green community production. Let' leave those distant customers to be served by another green community development someday.

I will include some statistics and calculations in Appendix E Table 3 for your convenience, but I have no good way of estimating the savings in fuel for all activities, so let's trust that the common sense way of doing business is feasible and effective in terms of fuel savings. I also think that much of the change in the mode of operation for the business community will be very transparent to most patrons except for possibly lower prices, so I don't foresee any real hardships being wrought on those involved in this new venture.

Keep in mind that this concept could be implemented in some fashion almost everywhere in the United States mainland, but at the risk of harping on Kansas let me show you how I determined the service area for the products. Kansas is a rectangular piece of territory, measuring 400 miles from east to west and 200 miles from north to south., and containing 105 counties. Most Kansas counties have the shape of a rectangle or a square, and many have about the same dimension, about 40 miles on a side. The estimated population for 2006 for these counties ranges from a mere 1,557 for Wallace county, "way out west in Kansas", and 470,895 in Sedgwick county, which includes the city of Wichita. Counties in the rural areas do well to muster 5,500 people and only a few are above 25,000 people. (1)

Quadrant wise, I calculated the estimated population for 2006 to be 95,962 for the NW quadrant, 217,601 for the SW quadrant, 1,475,050 for the NE quadrant and 975,732 for the SE quadrant. That is not many people by metropolitan standards, which serves to further support my claim that considerable savings of motor fuels can be had by doing a few things differently in the outback. I divided the state into four roughly equal land areas, called quadrants, and in my exercise I will sometimes refer to a service or sales area being a quadrant, which is 200 miles wide and 100 miles high. The distance from the center of a quadrant would be 100 miles to the nearest point on the east and west borders and 50 miles on the north and south borders. As the crow flies, the distance from the center would be a little over 110 miles to the farthest corner of the quadrant. Given these parameters, a small truck that uses some sort of biofuel could serve the area, which would avoid the expenditure of large amounts of diesel fuel.

The two western quadrants of the state are sparsely populated and most of the people live in the eastern quadrants, but as with public utility offerings, all patrons expect decent service within reasonable limits. This sparseness should serve to point out the wastefulness of doing things as usual because a lot of miles are covered to make a relatively few sales when the population density is sparse. One way to improve the sales efficiency would for people to bunch up in the more populated quadrants, but that is an emotional issue and is not likely to happen.

In some cases the optimum service area would be only a county, with about a twenty mile drive from a central point in the county to the far corner of the county. I will leave it to you to expand the scope and the effect of such fuel savings in your mind. Working with a small area permits me to use numbers that are easy to relate to, but if I were to do the math for the nation as a whole that could get speculative and confusing. Let it be said that these green concepts are intended to be adopted nationwide as the local situations permit. After all, a bucket of water is filled with drops of water, so the total effect of thousands of small developments would be a substantial savings of motor fuels and even energy to be used for domestic, commercial, and industrial purposes.

APPENDIX C

Housing That is Right For Your Needs and Tastes

Before going into detail on the various types of housing let me make a suggestion as to the type of construction. Modern multiunit complexes are typically stick built, meaning they have a wooden frame and roof members. Contrast this to the rugged fire and storm resistant dormitories on military installations and at colleges, where the building has a reinforced concrete frame, masonry exterior and interior walls, and concrete floors, ceilings, and stairwells. Such buildings are more expensive to construct than the stick built ones but they provide better shelter from storms than the wooden structures and are built for the long haul. As long as the foundations do not settle and crack the walls the reinforced concrete structures should last until the cows come home.

If the locale of the community is in a storm belt then I highly recommend the added investment of reinforced concrete construction, and I think the recent tragedy in little Greensburg, Kansas has made my point quite vividly. News reports of that little town showed only the concrete elevators and other sturdy structures standing. Need I say more about the virtue of storm resistance housing? I think this same principle could wisely be applied when making a hotel or other commercial buildings where large numbers of people might be injured by a storm strike if the building cannot stand the pressure.

In early Amana storms wrecked their factories more than once and it took a heavy toll on the financial resources of the community. Let's try to avoid that with Fort Hickok by doing it right the first time. My wife kids me because when I do a wooden project such as a picnic bench or a small building like a dog house I really lay the materials to it, and they are heavy as heck to carry around. That's alright honey, they are built to last.

Concrete and masonry structures seem to stand the test of time better than those made of wood, and the Greek and Roman ruins attest to that because they still stand while the little wooden structures those people used for domestic purposes are gone as if they have evaporated. Have you ever seen a big building in the downtown areas of cities fall victim to a tornado to the point it is flattened? Granted, the World Trade Center buildings collapsed but with a bit of help I fear. The buildings in the Amana Colonies are made of thick chunks of sandstone and are still standing and being used after 150 years or more. Once again, a tribute to those sagacious people and their building methods.

1. Boarding House

This type of housing would be designed for those needing only a suite or room in a boarding house and the units would be in a group setting with individual rooms and baths but with common dining areas, lounge, and laundries. A common kitchen might be part of a housing unit or could be centrally located, which is the way I lived while in the Air Force at Offutt AFB near Omaha. That arrangement also had the parking spaces for cars within an area bounded by the dorms, but the student quarters at the technical training base in Mississippi kept the cars on the perimeter of the dorm area and used a common kitchen that might be several dorms (barracks) away. These kitchens, or mess halls as the GIs called them, were designed to feed large numbers of students in a very short time so they could be on their way to school.

The boarding house would be a two story building made of storm and fire resistant materials and would consist of tenant rooms, a common lounge or recreation area, a common laundry room, and perhaps a common kitchen and dining room if a centrally located dining facility is not being used. Military dormitories often have a common mess hall (dining facility) amongst the housing units, and the mess hall building may also house a mailroom and a place to pick up fresh linen.

The individual units would have a large room that would include a bed, a table and chairs for dining, reading, and the like, a small kitchenette with a microwave, refrigerator, and sink, and a separate bathroom. A comfortable couch that could serve as a bed for guests and a nice lounge chair would also add to the comfort of the small unit. Storage space such as a wardrobe or dresser and a closet for the tenant's clothing would be provided. These units would resemble a large motel room that offered somewhat the same amenities, and the tenants would be expected to take their meals at a common dining facility, either in the same building or in another building.

A one car covered parking stall with secure storage in the rear of the stall would be provide for each rental unit, and extra open air parking would be included for visitors and for extra vehicles. Other outside amenities might include a balcony on the upper story and a small private patio on the lower floor for each room. The facility could also serve as an assisted living unit for those who were ambulatory and needed only a minimal amount of extra care such as house cleaning, laundry, and food preparation, and the cost would be much less than that of a typical assisted living facility.

Units with common walls, ceilings, and floors are more fuel efficient than those that are free standing because the heat gain or loss is shared from one unit to another. Additionally, those with interior hallways and stairwells reduce the heat gain or loss from coming and going to the individual units.

2. Assisted Living

Assisting living units are basically boarding house units where the tenants are able to get around on their own but want someone else to clean their suite, launder their clothing and bedding, and to provide the meals. The homes I have visited do offer a small amount of healthcare but the residents must be ambulatory enough to go to a common dining area for their meals. This type of housing is often cost prohibitive for those on a modest budget in the commercial homes but I think a basic setup could be offered at much the same rate as for a stay in a boarding house. Any special services desired by a given tenant could be obtained independently from an approved source. Using independent contractors means some additional income for residents in the community who want part or full time work providing those services. It is a type of caretaker arrangement.

These units could be simple and offering minimal assistance or could be comprehensive in scope. The local assisted living homes are very expensive and are lovely, with the residents living in efficiency suites and sharing a dining area and space for visitation and recreation. Those built to serve the aged and the handicapped should be single story and should provide features for the handicapped.

The individual rooms might be about the same size as those for the boarding house but simpler in the kitchen and dining areas.

3. Apartments

Second story apartments will be offered to those wanting that setting, and ground floor housing will be offered for families with small children and for old-sters. The idea is to reduce the injuries due to falling, which is a problem for small children and older people who don't get around so well. It may be a good idea to have the families live in certain areas and the adults in another. In other words, adult only areas. As we get older it seems our tolerance of the noisy chatter of youngsters gets a bit thin so adult only areas help to alleviate this problem.

Tenants with children should remain on the ground floor to avoid the hazards of the little ones falling down the stairs or throwing things off the balconies and hitting people down below. The individual units would be larger than a boarding house efficiency unit and would contain two bedrooms, an eat-in kitchen and a combination living-family room. Laundry would be done in a common area on the lower level, and a safe room for storm shelter would also be provided. Kids need a safe place to run and play so the entrances to the units would be to the inside, away from the street, and would lead to a common hallway that runs between the rooms. This arrangement also makes for better heating and cooling efficiency. The units would be furnished and would be rented on a bills paid basis except for the telephone service and something more than basic internet service, such as a digital cable internet service.

Single stall covered parking space would be provided as well as self storage space in the rear of each stall. Access to the complex would be via passageways since the entrances to the building would not face the street, and a small enclosed patio or yard on the street side of the building would be provided for the lower units. Such areas are nice to air out the little ones without having to worry about them running off down the street. The central or common mall area could be used for recreation, exercise, and even gardens if that were in demand.

4. Single Family Units

Single family units will be single story and will be shaped and built with con-servation of energy in mind. Space will be provided for communal activities to take place if the residents desire to do so. There should be an area for older residents who want to eat in a common dining room rather than prepare their own meals. This arrangement permits them to remain independent and out of a rest home, and would also provide common areas for recreation and lounging. Some may even want their own little garden plots, which is great as it helps them to feel self sufficient and provides them a way to socialize while leaning on the hoe! The units will be modest in size but well appointed with quality appliances and materials. A small house doesn't have to be a shoddily built house and it is false economy in the long run to build junk housing. The units should be durable enough for the long haul and could be either rented from an association or purchased.

Some houses would be two bedroom and others three bedroom, and would be sized and shaped to provide good energy efficiency. We Americans love our big homes but they can eat our lunches when the utility bills come due, so these homes will be tailored to the needs of those needing a given amount of space. For example, a couple starting out or a retired couple might not need or want a large three bedroom house so a smaller unit such as a two bedroom house would be offered.

The houses would all contain construction techniques, appliances and design concepts that promote the conservation of energy needed to operate the home, and would contain alternate heating provisions such as a pellet stove or a wood burning stove. Fireplaces are not particularly efficient so if a fireplace is included it should also include a slip in pellet stove or a slip in wood burning unit, both of which can be controlled to give better fuel efficiency than a blazing fireplace.

The units would have a living room dining room combination, and an eat-in kitchen. As much as I love our large formal dining room, modern lifestyles in general seldom utilize such amenities, so why bother? Spend the money on a space that will be better utilized. Attached garage and parking space is typically a feature of free standing homes, but an alternative is to have a separate parking garage or covered stalls for the cars and separate self storage space for the treasures that folks tend to accumulate.

If the homes are to be rented then rugged materials for the interiors should be considered, such as short loop carpet, interior wainscoting with wood paneling, and possibly ceramic tile in the wet areas such as the baths, laundry rooms, and the kitchen. Exterior walls need not be masonry veneer, but it is durable and impressive. Modern exterior materials such as the fiber-cement HardiPlank or HardyPanel, both being trademarks of James Hardie Industries, are very durable, contain a warranty for as long as 50 years, and are much less costly to buy and install than stone or brick masonry. I suggest that the exterior be highlighted in places with stone or masonry for emphasis and beauty, but I do not agree that either material make good insulation, for they don't. The R values for dense materials such as stone, brick, glass, and steel are much lower than for wood products and other materials that entrap air for added insulation.

There are provisions to ease the tax burden on owned homes and to ease the passage of the equity to an heir if the need be and I will try to address these topics in the communal living chapter of the book. The rental units would be the property of a business organization of a type that minimizes the tax load on the community. Taxes are needed for common functions such as fire and police protection and schools, just to name a few, but tax authorities can get carried away so the best approach is to shelter the tax liability right up front in the initial stages of a development.

5. Resort Housing

A portion of the community housing will be a resort type development and will be designed with the senior and more affluent people in mind. Provisions will be made for recreation areas and other amenities such as a clubhouse, which are usually not found in the more modest housing developments. However, the theme of conservation and affordability will be carried to these developments, as having extra money is no excuse for wasting energy. These will essentially be upscale apartment complexes much like those you might have visited in Branson, Missouri or in some other resort setting.

Resort developments are lovely places, as you may well know if you have ever visited them, and they provide recreational opportunities as well as nice housing units, which are basically small apartments that are time shared. The units are furnished, and include a small washer and dryer. Since they are in effect "owned" by the tenants, the amenities are more elaborate than with a furnished apartment where the landlord owns the furnishings. Maintenance and upgrade fees are commonly charged on an annual basis for the time sharing units, which is an arrangement where an individual purchases the right to use the unit for one or more weeks out of a year and can usually lend the privileges to relatives or friends.

I like the social amenities that are offered by the resorts, such as golf courses, tennis courts, swimming pools, and a clubroom for parties and gatherings. However, the time sharing schemes heavily favor those selling the time and the patron gets little for his rental dollar, which is basically what it is, a rental agreement. At Fort Hickok a user owned system would be my preference and the user or individual tenant would be left to share it with others. However, a staff would be available for maintenance and security, but the other services would be had by contracting with independent vendors. I think this is the fair way to do it because different people have different tastes for luxury and different budgets.

The resort units would emphasize common outside areas for rest and relaxation and would include the usual features of a resort community in a common courtyard, except for the golf course, which would be located elsewhere. Parking and storage would be provided for each unit as well as visitor parking, and the units would face inward with their back sides to the street, which means that special care should be taken to design a complex that has an attractive "back forty".

6. Hotel and Motel Rooms

Hotel or motel rooms will also be offered for visitors to the community, and will be designed to accommodate tour busses and gatherings for such events as class reunions, retreats, and small conventions as well as individuals or families. This part of the development should provide convenient access to the theater and the night spots of the town, and might even contain part of these facilities. With heads-up planning a package could be offered that included lodging and entertainment available in the area.

Some tenants want only a comfortable room in which to get a good night's sleep and others may want some more space, such as in a suite, if they are going to be staying for a few days. As with other group living units, the building should be fire and storm resistant and should have a shelter area in case of storms.

A hotel or motel that offers an area for gatherings, such as conventions or reunions, a nice quality dining facility, a lounge, and some exercise opportunities is a much more comfortable place than one that simply has a room and an ice machine. The rooms would have two queen sized beds, a full bath, a small refrigerator, and a microwave plus the usual tables and chairs for eating, lounging, and watching TV.

7. Senior and Handicapped Housing

Senior and handicapped housing is a necessary part of a community as we all wear out as time goes on, darn it, except for Millie our neighbor. She was ninety seven this month and is still growing her garden and drives her little Chevy Blazer. Still truckin', as they say in the outback. The senior housing facilities in Abilene vary from very modest (and poorly made) single story units that sit side by side in a row, to three story facilities that are in top shape and are very attractive. The individual spaces are typically a small efficiency apartment with a kitchenette, full bath with a walk in shower, a combination living dining room, and one bedroom. The ground floor units have access to a small patio and the upper floor units have a balcony. Multi story compounds for seniors and handicapped people usually have elevators, but those on the upper floors are left to use the stairways in case of fire, so that may pose a problem for tenants that do not walk well. The living quarters share a common hallway, which makes for better heating and cooling efficiency of the unit because it reduces heat loss and gain from going in an out so many doors that pass to the outside environment.

The projects I have seen are lovely and appropriate but none of them have made provision for alternate energy. Perhaps some of the older people cannot work with a pellet stove because the bags of pellets weigh 40 pounds, but that could be handled by an assistant in the complex. The better housing units here in Abilene seem to have plenty of staff so that may be very possible. If the 40 pound bags of pellets could be subdivided in to smaller bags then the older tenants could probably handle them without assistance.

The individual units are typically small but the complexes try to provide a modicum of entertainment and elbow room outside for flowers and lawn chairs in the shade. Perhaps seniors and handicapped people don't expect or need much in the way of group amenities but why not give them a break. After all they have paid their dues in life and deserve some special attention.

There is a lovely group home in the nearby countryside that provides a little chapel, housekeeping, laundry service, cafeteria meals, craft and recreation rooms, and a pond of ducks to laugh at as they waddle around and make a terrible mess of things. The whole thing is subsidized by a foundation and the cost to stay there is very reasonable, given the attention the tenants receive. The development is known as the Brown Home and is located in a lovely hilly country setting that once was a picnic and play ground for visitors. Those that can still drive have their cars there and are free to come and go on their own as they choose. Some of the residents do volunteer work in town, so the place is home to a number of older folks who are still contributing to the ambience of the city of Abilene. (1)

8. Rest Homes

Rest homes are a fact of life for many, sad to say, and some of those in Fort Hickok may need a rest home stay. Hopefully some quality care can be offered at reasonable prices but the business is labor intensive and tends to be expensive. Fort Hickok is likely to need one or more rest homes and they should be included in the housing mix. A relatively new idea, adult daycare, is an alternative to staying full time in a rest home for some. This is a nice service for those who are taking care of someone but need to work outside the home for the day. The patient can be taken to a facility where the usual tasks of bathing, feeding, and such are done while the caregiver works, and then is brought back home in the evening.

Rest homes and day care homes use specially sized and equipped rooms so the staff can come to a room and work with the patient, and they typically use a common dining room, or more than one if the center is using a new concept of rest home care and dining, called a community concept, where different wings or groups of the home dine together as a group rather than going to a large common dining room. This supposedly provides a better sense of belonging to the residents than being one of a great number of strangers in a large and noisy dining area.

Group homes such as these need to be well constructed and should be storm and fire resistant, and for the convenience of the staff and the wheel chair residents it is best the facility is single story. Some rest homes have wings spread out in a star shaped pattern with rooms on each side of a hallway that runs the length of the wing. This supposedly permits more light and air to reach the rooms than with a closed up floor plan but it strikes me as a fuelish way to heat and cool a building. A closed plan will serve to reduce utility costs and can be creatively designed to provide interior courtyards, possibly closed at the top, for the residents to enjoy.

APPENDIX D
Methods For Making Biofuels

The processes used to make ethanol fuel from grains and biodiesel from animal fats and oilseeds are commonplace and might be something that could be done at Fort Hickok. However, there is a limit to the amount of grain that America can devote to making ethanol without impacting the other uses for those grains, and there is also a limit to the amount of feedstock that is available for biodiesel production. The following methods are being tried at various levels in the United States with some being fully at a commercial production level and others being at more of a prototype level. The types that are right for Fort Hickok will depend upon the climate, rainfall, and terrain of the locale in which the community is located. Thankfully these processes do not depend upon large masses of grains because many states do not grow that much grain but have a bounteous assortment of trees, shrubs and other natural plant growths. Some of the processes can even utilize municipal waste or old tires for feedstock so in that respect are more conservative of energy than having to use grains for feedstocks because less effort and input is needed to produce the feedstock.

1. Cellulosic Ethanol (1)

The term cellulosic refers to the cellulose in plant materials, and the ethanol derived by the processing of the cellulose is the same at that derived from fermentation and distillation of grains. Using plant wastes, or more generally biomass, the processes used to make alcohol this way are not dependent upon grain for feedstocks, which makes the method portable in that it can be used in areas that are not rich in grains. However, if large amounts of water are needed for the processing, this is seen as a weakness in the states where water is becoming scarce, which also may be a limiting factor for producing ethanol by the fermentation and distillation methods now being used.

The online encyclopedia, Wikipedia, has an article on cellulosic ethanol that explains the processes that can be used to make ethanol in this manner. Gasification is an alternative to the cellulolytic method, the one that requires lots of water, and is also called synthesis gas fermentation or catalysis (e.g. the Fischer-Tropsch process), and it too seems to require water as it involves a fermentation step. There may be hope on the horizon because a company called Syngas International Corp. proposes to use catalytic conversion rather than fermentation to produce ethanol. (9)

2. Methanol Production (2)

For some reason there has been little interest in making methanol for motor fuels but the potential is tremendous, some 455.7 billion gallons of methanol, which is equivalent in energy to 259.75 billion gallons of gasoline, gasoline having more energy per gallon of the two. Even then, this is four times the United States consumption of 63 billion gallons of gasoline in 2003. Two objections to using methanol is that it is caustic to the fuel systems, and water separation with a methanol-gasoline mixture is more of a problem than with ethanol. Both objections can be overcome so what is the holdup? Could it be that because most of today's methanol is made from natural gas, a non-renewable feedstock? Methanol can also be fed to a fuel cell that is designed to utilize the hydrogen atoms in the methanol compound, so this too would seem to be a safe way to distribute hydrogen, by transporting the liquid methanol rather than the gaseous hydrogen.

3. Gasification (3)

Gasification is a process that can be used to convert organic materials (biomass) into syngas, which is about 85% hydrogen. So once again we see a way to get off the petroleum barrel, as hydrogen can be used as a motor fuel directly or as a feedstock for fuel cells to produce electrical power. Additional processing can be used to make methanol from the syngas, so this process is versatile rather than being restricted to producing only one usable product.

The Department Of Energy article speaks of coal gasification but states that the process could also be applied to other biomasses to produce output fuels with low emissions of sulfur and nitrogen oxides. If oxygen is used in the gasification process rather than air then carbon dioxide is emitted as a concentrated gas stream, which is easier to sequester than when using air. In my opinion, if electrolysis were to be used to produce hydrogen, then the resulting byproduct of oxygen could be fed to this process.

4. Butanol Production (4)

Butanol is a type of alcohol that can be used as a motor fuel, and can be produced from biomass as well as from fossil fuels. Production of this substance also involves fermentation, the old bit of changing the starches to sugars so an alcohol product can made and finally distilled from the water. An advantage of creating butanol is that it also can be made from plant materials in addition to seed grains, so this method would help to take the load off the grain supply. Areas east of the Grain Belt such as Missouri, Arkansas, and eastern Oklahoma have carpets of trees rather than carpets of grain so these areas would be well positioned to make butanol from plant materials. Butanol has a number of desirable characteristics and rivals gasoline in energy content per gallon. The Wikipedia encyclopedia has a good write up on butanol.

5. Hydrolysis (5)

Hydrolysis is a chemical decomposition process that is used to split chemical bonds of substances. Wow, you can sleep well tonight wondering what in heck I have just said. Citric acid can also be produced by the process, and has a variety of uses. In general, a *saccharification* process is used to convert the material to glucose or starch and then the glucose and starch is converted to an alcohol fuel. This process could also utilize plant residues and other organic materials as feedstocks, but it does require additional water to be added.

6. Pyrolysis (6)

It sounds like science fiction, sorta hot and scary? Well, it could be if you were the feedstock being processed. Pyrolysis liquefies solid biomass or organic materials, producing an oil similar to distillate (like kerosene or diesel maybe), which can be used to power diesel engines and to fire boilers, so here might be a partial alternative to petroleum based diesel fuel and thank Goodness because we simply could never raise enough oilseeds or animal fat to replace the current supply of petroleum based diesel fuel.

In a process called flash pyrolysis approximately 75% of the input feedstock is converted to oil, and feedstocks include bagasse, palm residues, rice husks, straw, dried sludge (from municipal sewer plants), pine wood, beech wood, oak wood, switch grass, and poplar wood. Poplar has the virtue of being a fast growing tree so it would be a good candidate for a woodlot intended as feedstock for the pyrolysis process. My my, look at those "p's". Peter Piper Picked a Peck of Pickled Peppers. There, now we can move on.

7. Anaerobic Digestion (7)

This is a low technology process that can produce methane gas from manure, sewage, sludge, and wet organic materials. The output is rich in methane, which can be used in heating as well as for a motor fuel. Methane, which is essentially natural gas, has been used for years as a motor fuel in fleets of busses and by corporate vehicles.

8. Partial Oxidation (8)

Here is the companion process for making all that methanol at the rate of 186 gallons per ton of feedstock, and a variety of feedstocks can be used, making this process versatile enough to be used in various locales with differing types of biomass input.

I hope by now that I have convinced you that there is life after the oil patch and that we need to get on with these promising methods of producing an alternative to petroleum based fuels and natural gas. Two virtues come to mind for Fort Hickok doing this. Number one, the processes use renewable feedstocks that are commonplace, and number two, the processes are green in that they use carbons that exist in live form or have resulted from plant growth that consumes carbon dioxide. It is thinking like this that helps to keep the carbon equation in balance and it is thinking like this that can improve our self sufficiency as a nation in regards to motor fuels.

APPENDIX E

Tables

TABLE 1. WATER CONTENT OF SELECTED FOODS

Food Item	Percent Water	
Carrots, Raw	87.8	
Lettuce, Iceberg, Raw	95.2	*
Grapes, Raw, American Type	81.3	*
Mayonnaise Type Salad Dressing	39.9	
Margarine Spread, Tub, Unspecific	37.0	
Catsup	66.6	
Wheat Flour, Enriched, All Purpose	11.9	
Cornmeal, Degermed, Enriched, Yellow	11.6	
Navy Bean, Raw	12.4	
Spaghetti, Dry, Enriched	10.3	
Macaroni, Dry, Enriched	10.3	
Bread, White, Common	36.7	
Grapefruit, White	90.5	*
Orange, Raw, All Varieties	86.8	*

Food Item	Percent Water
Ground Beef, Lean, Raw	60.2
Egg, Chicken, Whole, Raw	75.3
Bacon, Cured, Raw	31.6
Ham, Cured, Lean, Unheated	68.3
Fresh Pork Sausage,	44.6
Beef Arm Pot Roast, All, Lean, Raw	72.9
Chicken Breast, No Skin, Raw	74.8
Farm Catfish, Raw	75.4
2% Lowfat Milk, Fluid	89.2
Ice Cream, Vanilla	61.0
Butter, Regular	15.9
Pasteurized Processed American Cheese	43.2
Cheddar Cheese	36.8
Jams and Preserves	34.5
Apple, Raw, With Skin	83.9
Potatoes, With Skin, Raw	83.3
Tomato, Red Ripe, Raw	93.8
Strawberries, Raw	91.6
Peach, Raw	87.7
Melon, Cantaloupe, Raw	91.5
Watermelon, Raw	89.8
Onion, Raw	89.0

The information in the above table was derived using an online Nutrition Analysis Tool furnished by the University of Illinois Extension Service. I printed these figures in October of 2004 but could not find the tool recently (May 2007). Not to fear as the data compares favorably today (May 2007) with that by Walton Feed at www.waltonfeed.com/self/h2ocont.html.

* The items marked with and asterisk would have to be shipped into the region unless the climate permitted them to be grown. Southern Texas, Florida, and California are good locales for these crops.

TABLE 2. CONSUMPTION OF SELECTED FOOD ITEMS

Food Item	Needed Per Person	Data Year	Needed For 1,500 people	Needed For 6,000 People
Strawberries (1)	6.1 lb	2006	9,150	36,600
Sweet Corn, Fresh	8.6 lb	"	12,900	51,600
Cucumbers	9.6 lb	"	14,400	57,600
Watermelon	15.8 lb	"	23,700	94,800
Cantaloupe	9.3 lb	"	13,950	55,800
Onions, Fresh	19.8 lb	"	29,700	118,800
Bell Peppers	7.7 lb	"	11,550	46,200
Tomatoes, Fresh	19.8 lb	"	29,700	118,800
Potatoes, Fresh	42.0 lb	"	63,000	252,000
Sweet Potatoes	4.6 lb	"	6,900	27,600
Mushrooms	4.0 lb	"	6,000	24,000
Meat, Boneless, Trimmed, Edible Weight				
Beef	62.4 lb	2005	93,600	374,400
Chicken	60.4 lb	"	90,600	362.400
Pork	46.5 lb	"	69,750	279,000
Eggs In Shell	175 ea	2005	262,500	1,050,000
Beer	21.3 gal	"	31,950	127,800
Wine	2.4 gal	"	3,600	14,400
Distilled Spirits	1.4 gal	"	2,100	8,400
Regular Soft Drinks	35.5 gal	"	53,250	213,000
Diet Soft Drinks	16.0 gal	"	24,000	96,000
White and Whole Wheat Flour	122.3 lb	"	183,450	733,800
Durum (Pastas)	11.8 lb	"	17,700	70,800
Corn Products				
Flour And Meal	18.8 lb	"	28,200	112,800
Hominy And Grits	8.1 lb	"	12,150	48,600
Starch	4.5 lb	"	6,750	27,000

Food Item	Needed Per Person	Data Year	Needed For 1,500 people	Needed For 6,000 People
Oat Products	4.6 lb	"	6,900	27,600
Barley Products	.69 lb	"	1,035	4,140
Whole Milk				
Plain	6.6 gal	"	9,900	39,600
Flavored	0.30 gal	"	450	1,800
Fluid Milk Products	21.0 gal	"	31,500	126,000
Yogurt	1.0 gal	"	1,500	6,000
Cheese Products Total	31.4 lb	"	47,100	188,400
American Cheese	12.7 lb	"	19,050	76,200
Other Cheese	18.7 lb	"	28,050	112,200
Cheddar	10.1 lb	"	15,150	60,600
Mozzarella	10.2 lb	"	15,300	61,200
Swiss	1.2 lb	"	1,800	7,200
Cream and Neufchatel	2.3 lb	"	3,450	13,800
Cottage Cheese				
Regular	1.3 lb	"	1,950	7,800
Low Fat	1.3 lb	"	1,950	7,800

Data from Economic Research Service, USDA unless otherwise noted
(1) Strawberry data from Agricultural Marketing Resource Center, University of California (Created March 2006) *Commodity Profile: Strawberries*

TABLE 3. FOOD NEEDED FOR 100,000 PEOPLE

Food Item	Per Person	1,500	6,000	100,000
Strawberries (1)	6.1 lb	9,150	36,600	610,000
Sweet Corn, Fresh	8.6 lb	12,900	51,600	860,000
Cucumbers	9.6 lb	14,400	57,600	960.000
Watermelon	15.8 lb	23,700	94,800	1,580,000
Cantaloupe	9.3 lb	13,950	55,800	930,000
Onions, Fresh	19.8 lb	29,700	118,800	1,980,000
Bell Peppers	7.7 lb	11,550	46,200	770,000
Tomatoes, Fresh	19.8 lb	29,700	118,800	1,980,000
Potatoes, Fresh	42.0 lb	63,000	252,000	4,200,000
Sweet Potatoes	4.6 lb	6,900	27,600	460,000
Mushrooms	4.0 lb	6,000	24,000	400,000

Meat, Boneless, Trimmed, Edible Weight

Food Item	Per Person	1,500	6,000	100,000
Beef	62.4 lb	93,600	374,400	6,240,000
Chicken	60.4 lb	90,600	362.400	6,040,000
Pork	46.5 lb	69,750	279,000	4,650,000
Eggs In Shell	175 ea	262,500	1,050,000	17,500,000
Beer	21.3 gal	31,950	127,800	21,300
Wine	2.4 gal	3,600	14,400	2,400
Distilled Spirits	1.4 gal	2,100	8,400	1,400
Regular Soft Drinks	35.5 gal	53,250	213,000	3,550,000
Diet Soft Drinks	16.0 gal	24,000	96,000	1,600,000
White and Whole Wheat Flour	122.3 lb	183,450	733,800 (2)	12,230,000
Durum (Pastas)	11.8 lb	17,700	70,800	1,180,000
Corn Products				
Flour And Meal	18.8 lb	28,200	112,800	1,880,000
Hominy And Grits	8.1 lb	12,150	48,600	810,000
Starch	4.5 lb	6,750	27,000	450,000

Food Item	Per Person	1,500	6,000	100,000
Oat Products (3)	4.6 lb	6,900	27,600	460,000
Barley Products (4)	.69 lb	1,035	4,140	69,000
Whole Milk				
Plain	6.6 gal	9,900	39,600	660,000
Flavored	0.30 gal	450	1,800	30,000
Fluid Milk Products	21.0 gal	31,500	126,000	2,100,000
Yogurt	1.0 gal	1,500	6,000	100,000
Cheese Products Total	31.4 lb	47,100	188,400	3,140,000
American Cheese	12.7 lb	19,050	76,200	1,270,000
Other Cheese	18.7 lb	28,050	112,200	1,870,000
Cheddar	10.1 lb	15,150	60,600	1,010,000
Mozzarella	10.2 lb	15,300	61,200	1,020,000
Swiss	1.2 lb	1,800	7,200	120,000
Cream and Neufchatel	2.3 lb	3,450	13,800	230,000
Cottage Cheese				
Regular	1.3 lb	1,950	7,800	130,000
Low Fat	1.3 lb	1,950	7,800	130,000

Data from Economic Research Service, USDA unless otherwise calculated

(1) Strawberry data from Agricultural Marketing Resource Center, University of California (Created March 2006) *Commodity Profile: Strawberries* (continued)

(2) Wheat flour is assumed to be 83% of total wheat grain weight

(3) Oat groat is assumed to be 75% of oat grain weight

(4) Barley groat is assumed to be 75% of barley grain weight http://www.fda.gov/ohrms/dockets/dailys/04/nov04/113004/04p-0512-cp00001-03-Appendix-01-vol1.pdf

TABLE 4. ACREAGE AND LIVESTOCK NEEDED FOR DEMAND (1) (6)

Food Item	Needed For 1,500	Needed For 6,000	Yield Per Ac. /Animal	Ac. /Animals 1,500	6,000
Strawberries	9,150	36,600	44,500 lb	0.21	0.82
Sweet Corn, Fresh	12,900	51,600	11,800 lb	1.09	437
Cucumbers	14,400	57,600	17,183 lb	0.83	3.35
Watermelon	23,700	94,800	34,000 lb	0.70	2.78
Cantaloupe	13,950	55,800	18,518 lb	0.75 (7)	3.01
Onions, Fresh	29,700	118,800	44,000 lb	0.675	2.7
Bell Peppers	11,550	46,200	28,400 lb	0.40	1.61
Tomatoes, Fresh	29,700	118,800	29,900 lb	1.0	4.0
Potatoes, Fresh	63,000	252,000	38,200 lb	1.65	6.6
Sweet Potatoes	6,900	27,600	17,600 lb	0.39	1.57
Mushrooms	6,000	24,000	5.73 lb/sq ft	1047 sq ft	4188

Meat, Boneless, Trimmed, Edible Weight

Beef	93,600	374,400	715 lb	1.31	524
Chicken	90,600	362.400	3 lb	30,200	120,800
Pork	69,750	279,000	145 lb	481	1,924
Eggs In Shell	262,500 ea	1,050,000	5 eggs/wk	1,010	4,040
White and Whole Wheat Flour	183,450	733,800	2610 lb	85 (3)	340
Durum (Pastas)	17,700	70,800	2280 lb	9.4	37.4
Corn Products					
Flour And Meal	28,200	112,800	8,982 lb	3.13	12.55
Hominy And Grits	12,150	48,600	"	1.35	5.41
Starch	6,750	27,000	"	0.75	3.00

Food Item	Needed For 1,500	Needed For 6,000	Yield Per Ac. /Animal	Ac. /Animals 1,500	6,000
Oat Products	6,900	27,600	2,070 lb	4.43 (4)	17.73
Barley Products	1,035	4,140	3,331 lb	0.41 (5)	1.65
Whole Milk					
Plain	9,900 gal	39,600	22,000 lb/yr	3.74 (2)	15
Flavored	450 "	1,800	"	0.17	0.68
Fluid Milk Products	31,500 "	126,000	"	12	47
Yogurt	1,500 "	6,000	"	0.57	2.27
Cheese Products Total	47,100 "	188,400	"	17.83	71.33
American Cheese	19,050 "	76,200	"	7.21	28.85
Other Cheese	28,050 "	112,200	"	10.62	42.48
Cheddar	15,150 "	60,600	"	5.73	22.94
Mozzarella	15,300 "	61,200	"	5.24	20.95
Swiss	1,800 "	7,200	"	0.68	2.72
Cream and Neufchatel	3,450 "	13,800	"	1306	5.22
Cottage Cheese					
Regular	1,950 "	7,800	"	0.73	2.95
Low Fat	1,950 "	7,800	"	0.73	2.95

(1) Demand is in pounds unless otherwise noted.

(2) 22,000 pounds per year is a conservative estimate from Wikipedia article on Holstein

(3) Assume wheat flour is 83% of grain weight, so acreage is increased by 1.20

(4) Oat groat is 75% of total oat grain weight, so acreage is increased by 1.33%

(5) Barley groat is 75% of total barley grain weight so acreage is increased by 1.33%

(6) Crop year for grains is 2004

(7) Yield per acre is conservative, for Indiana in 2004

TABLE 5. WHOLE MILK CONVERSION

To make one pound of	Requires
Butter	21.2 lbs. Whole Milk
Whole Milk Cheese	10.0 lbs. " "
Evaporated Milk	2.1 lbs. " "
Condensed Milk	2.3 lbs. " "
Whole Milk Powder	7.4 lbs. " "
Powdered Cream	13.5 lbs " "
Ice Cream (1 gal.)	12.0 lbs. " " (a)
Cottage Cheese (Dry Curd Basis)	7.25 lbs. Skim Milk
Non-fat Dry Milk	11.0 lbs. Skim Milk

(a) 15 pounds when including butter and concentrated milks

Data reproduced from a table titled Whole Milk Conversion in *Value-Added Processing Feasibility* Report by Jay Hammarlund, Cooperative Development Specialist (July 2003) of the Kansas Department Of Commerce, Agriculture Marketing Division

APPENDIX F

Small Community Wastewater Treatment

Treatment of wastewater is a necessary part of having a community, or even a sole residence for that matter. There are a number of options open to the small community, which is expertly pointed out in *Individual Homeowner & Small Wastewater Treatment and Disposal Options*, an article on the internet by Doley and Kerns (1). I don't want to get too deep in this mucky topic for fear of never surfacing to bless you again with my hokey vernacular, so I will just touch on the parts that appeal to me relative to the need at Fort Hickok.

At first it may be most affordable if each residence had its own septic tank and cesspool or lateral to handle the household waste, as it is relatively inexpensive and trouble free to do it that way. However, once a community builds up the individual systems tend to take up a lot of land, depending upon the ability of the soil to absorb the affluent, which is called the percolation rate, and I have seen areas where there can be no more than one septic system per three acres of land area. That restriction would not permit a dense development where dozens of people might be living on three acres or even less, as would be the case with a rest home or some other kind of group home.

There are stages involved in wastewater treatment, which is to be considered when designing a system for a community.

1. Primary Treatment is the separation of solids and particulate matter from the wastewater and it is usually done in a settling chamber, which is a physical process.

2. Secondary Treatment is the reduction of organic compounds in the wastewater, which is a biological process where the bacteria digest the compounds.

3. Tertiary Treatment is any stage where the water is polished through a filtering process or where nutrients are otherwise removed. Polishing refers to the removal of very fine particles from the water, and the nutrients such as nitrogen and phosphorus also need to be removed so they do not pollute lakes and streams. These nutrients feed the algae and cause problems in the bodies of water.

4. Disinfection is the removal of any possibly harmful pathogens before the effluent is discharged. One way of doing this is to chlorinate the effluent to kill the bugs and then to remove the chlorine before letting it loose to a stream.

I told you it may get sticky, and am I glad I suffered through the heavy stuff above. Now, let's move on to some easy pickin' (Oh My!). The article points out a number of ways to process the wastewater but one struck me as being right for Fort Hickok. So you see there is a method to my madness. I'll bet you suspected I had an ulterior motive for dabbling in the juicy stuff, but that shouldn't surprise you by now. The subtopic *Cluster Systems and Small Centralized Community Systems* deals with waste systems that can handle the needs of a small community, perhaps as much as 10,000 people, so such a size would allow for some growth at Fort Hickok.

The Sequencing Batch Reactor method appeals to me for at least three reasons. Firstly, the effluent to be discharged from such a system is very low in organic compounds, and secondly, a relatively small area of land is needed. A feature that really caught my eye is that the system is modular in that it can be easily expanded as needed to handle more customers. And last but not least, such a process results in sludge, which is a crumbly smelly substance that contains a low amount of nitrogen, about five percent. This magical stuff can be incorporated into soil to increase its workability. I might be reluctant to use sludge in a soil that was used to grow fruits and vegetables but using it on farm land for cereal grains and the like should pose no problem. One might seek the advice of the health department before doing so.

It is for sure, sludge has a magical quality when used on lawns. I once used it on my lawn and when I told the kids what it was, for some reason they stopped running across my grass! Just goes to show you, kids don't mind making messes but they don't particularly like to play in messes. Did I make messes when I was a kid? Just ask mom.

My boyhood home was in a small town of 300 people, and folks used septic tanks and cesspools or laterals to handle their household waste. Our cesspool started to overflow after the flood in 1951 so my dad, being a good neighbor, pumped the effluent on the neighbor's potato patch, which was a big one. The effluent killed a bunch of the potatoes instead of fertilizing them as my dad expected it to do, and little Annie Stirn was mad as a wet hen. Such is life in a small town with farmers around.

Appendix G

Aermotor Windmill (1)

Aermotor windmills are still being manufactured, and web site www.
aermotorwindmill.com (accessed in June 2007) has information on the capacity
of the different sizes of mills that are available from the Aermotor people. The
capacity of a mill is determined by three things, the velocity and duration of the
wind (Oh Really?), the diameter of the wheel, or fan as I call it, and the diameter
of the pump cylinder, which is the business end of the contraption down in the
well water. Another feature has a bearing on the capacity, and that is whether or
not the mill is set for a long stroke or a short stroke, which is something I wasn't
aware of when I used to help my dad pull the cylinder and replace the leather gas-
kets.

The Aermotor article speaks of different sizes of mills and is speaking of the
diameter of the wheel. For example an A size mill has an eight foot diameter
wheel, a B size a ten foot diameter wheel, and an E size a fourteen foot diameter
wheel. Cylinder size refers to the diameter of the pump cylinder, which is usually
made of iron and has a piston in it that lifts the water on each up stroke of the
mill head. Think of the mill head as the engine because it reciprocates in a way to
move the sucker rod up and down to work the cylinder, which is attached to it.
Aermotor calls it the motor, no kiddin'!

Keep in mind that some of what I say is based on the Aermotor data and some
is coming from my hazy recollection of 50 years ago when I used to work (play?)
with my dad's windmill. His mill was in a valley because that was where the water
was closest to the surface, and the well was hand dug, about 30 feet deep and
maybe 6 feet across at the top. The water was usually cool, even in the heat of
summer, and the cows loved to stand at the tank, slurp that cool well water, and
swat a few horse flies with their tails.

In time my father's well got so it wouldn't produce much in the dog days of summer because silt had stopped up the walls to the point that little water flowed into the well when it was needed most, darn it. He remedied the situation by hooking up to the rural water district line, which ran by the front of the property, so he could have a dependable supply of water for his cattle all year long. Rural water districts plus the advent of electrical power in the outback, the Rural Electrification Administration (REA) for some, spelled the downfall of many graceful old windmills that had served faithfully for decades, through drought and blizzard and wind storm. The REA still exists and can be found on the internet by using a search engine such as Google, which is a proprietary software product, and a very powerful one at that.

A deep well, in fact most modern wells, would be dug with a machine and would be a relatively small bore (diameter) with a metal or plastic casing running from the top down to the zone where the water comes in to the well. It seems to me that a deep well would take more power to bring the water to the surface than a shallow well would, for this reason. The cylinder, which is down in the water near the bottom of the hole, is fitted with a piston and valves so that on the up stroke it lifts (pushes up) the amount of water that has been drawn into it during the down stroke and the brief time when it is resting on the bottom of the stroke. So what we have is a column of water as high as the well is deep, plus some more if the water is to be lifted above ground level. This is common with windmills, to lift the water several feet into the air above the ground level so it can flow into a reservoir, and you may have seen the old wooden tanks in wild west movies where the pump was storing water for the steam engines on the trains. My theory is that the taller the column of water to be lifted by the cylinder, the harder the motor on the top of the tower has to work. Farmers also used various types of reservoirs, and they ranged from simple barrel like tanks to large ones built of concrete, the latter being what I see being used at Fort Hickok.

Windmills turn slowly when compared to wind driven power generators, and they have a tremendous amount of torque, even at low speeds. The speed of the wheel depends first upon the wind velocity, but it can be altered by using a lever to pull the wheel partly out of the path of the wind. This lever, which is mounted on one of the legs of the mill tower, is connected to the mill head by a long heavy wire or chain, and by moving this lever up or down one can change the speed of the fan, even stop it, or feather it, somewhat like feathering an airplane propeller in that the vanes of the wheel (the propeller) are moved so that they are more nearly parallel to the wind direction and therefore develop less turning force on the axle of the wheel. Shutting the mill down consists of pulling the tail up close to the wheel, which makes both the pieces parallel to the wind direction. If my memory serves me correctly I believe that some windmills also had a mechanical brake that was engaged when the mill lever shut the mill down, and this served to further protect the wheel against wind gusts shifting direction in a wind storm.

It seems that we tend to take things for granted when we are young, even a clanky old windmill, but the engineering behind the dumb thing is very clever. Mills have evolved from the early wooden towers with wooden fans to the modern galvanized steel towers and fans. At one time it was a real status symbol for a farmer to have a shiny new Aermotor or some other brand of mill on his place, and it may still be because mills do not come cheap any more than gasoline does these days. However, the dependable old wind driven mill doesn't pollute, and simply does its job decade after decade with only an occasional amount of maintenance or repair. The crank case of the motor (atop the tower) may need some oil added once in a blue moon and the pump cylinder leathers (at the bottom of the well) may need to be changed once in a while, particularly if the pump is sucking sand, but aside from that the darned things make a wonderful conversation piece, and the old cows love to use the tower legs for scratchin' poles.

You may have seen an old windmill on display at a historical town site. Great, but don't underestimate the power of these clunky old engines because they have helped to put the meat on America's tables for over a century now and they can still do their duty at Fort Hickok. Use Google (2) to look up "Aermotor windmills" or "Dempster windmills" on the internet and you can have a pleasant trip down memory lane. Dempster is located in Beatrice, Nebraska and I pass by the factory as I go thru there on the way to Lincoln, Nebraska. There may be other mill companies that still produce mills but many have gone bye bye, sad to say.

The Aermotor web site lists the Average Water Needs for farm animals and for the home, so you might want to check it out. Here are a few entries from the table.

Average Water Needs	Gallons
Milking cow, per day	35
Dry cow or steer, per day	15
Horse, per day	12
Hog, per day	4
Chickens (per 100) per day	6
Flush toilet, per filling	2-7
Kitchen sink, per day	20

Well now, the kitchen sink usage can be reduced dramatically. Just eat out more often! Duh? Some of the numbers may be obsolete due to design changes in home appliances, but the old milk cow you can be sure still needs her water each day. I think a few big windmills in the Fort Hickok area could supply much of the water for livestock and some for domestic uses such as watering grass, car washes, and other applications where sterile water is not needed.

Being a farm kid I find something nostalgic about an old mill moaning and clanking, especially when the wind is just enough to turn it, and the expectation of another shot of cold water from the mouth of the cast iron pump still gets me, even at 68 years old. A little rust and sand from the cylinder deep down in the well also helps to improve that taste of that cool drink, almost as good as a cold beer. Well, almost I said.

APPENDIX H

Wheat Milling

Wheat is king in Kansas and in a few weeks from now there will be oceans of it waving in the wind, that is, what has not been flooded out or frozen out. This year hasn't turned out to be the best of years for some of the wheat farmers but we usually have mountains of it lying around waiting to be shipped out of the area for processing. Web site www.namamillers.org/ ci_products_wheat_mill.html (1) has a good article on the wheat milling process and another site, www.wheatmania. com/allaboutwheat/wheatfacts/howwheatbecomesflour.htm (2), gives essentially the same description of the process but includes an annotated diagram of the process that makes it easier to visualize. Let me touch on some of the major points of this process, and keep in mind that different mills may have a slightly different routine depending on the type of flour to be made. Appendix K lists the web address of a small independent mill in Kansas and shows the various products being offered by this small mill. My wife bought some Hudson Cream short patent flour just yesterday.

The wheat is tested for quality and type to determine the type of flour it is best used for. After a variety of cleaning steps the grain is tempered or moistened, which permits the husk or bran (the light tan skin) of the berry to removed from the starch (the endosperm) and the germ. In time the germ will also be separated from the starch, which is the flour, because if left in the flour the oils of the germ will not let it keep well.

Mills are designed to use the force of gravity to pull the wheat in its various states of process down thru the grinders or rollers, the sifters, and such, so that may explain why the older ones were many stories high. I used to go by the big mills in Salina, Kansas at night and see the fluorescent lights burning high in the mills above the street. Those mills apparently ran non-stop until the supply of grain had been consumed.

A number of enrichment ingredients such as iron, niacin, and riboflavin can be added as desired to make *enriched flour*, and a bleaching process improves the baking quality of flour, as it was noted years ago that freshly made flour did not work as well as aged flour, so the bleaching step provides the aging during the milling process. Now I know why there is *unbleached* or *bleached* written on the bags of flour. Of course you bakers knew that all along. My momma didn't tell me that because I was too busy slopping the hogs with the *shorts*.

Flour constitutes about 75% of the input grain, and the remainder is separated out as feed for livestock, some being middlings or shorts, some being bran, and some being wheat germ, which can also be used for human consumption. A bit of middling is a tiny piece of the bran that is stuck to a speck of endosperm or starch (the flour) and an alternate name for it in the trade is shorts. Hogs like a concoction of shorts and warm whey from the cheese factory, and the old milk cows love the bran, which is a bit like the bran flakes you may eat for breakfast except that it is soft and is not toasted.

One more tidbit that you will need to sleep well tonight. There is such a thing as long-patent and short-patent flour, so let me try to pass on what I learned about that terminology. The folks at the Stafford County Mills Co. tell me this about short-patent. This type of flour results from more sifting and crushing than the long-patent flour, and has a lot of the heart of the endosperm in it. See web site www.hudsoncream.com (3) for more details on this type of flour. It is used for pastries and such. Long-patent has a greater portion of the total endosperm in it and is used for bread making, amongst other things. It apparently is a bit coarser than the finer short-patent flour.

The term *patent* refers to the amount of endosperm used to make the flour, and the following web site gives a nice explanation of the various patents and what they are best used for. See http://www.aces.edu/dept/extcomm/newspaper/dec3c01.html (4), and I thank the Alabama Cooperative Extension System for this input.

Please read the articles in web sites I have presented above, and you might learn something new about what goes into making the staff of life. I know I sure did, and I have taken flour for granted all my life, as I grew up in a wheat farming family here in central Kansas.

Appendix I

Corn Milling

The North American Miller' Association also gives a description of milling corn, and the information can be found at www.namamillers.org/prd_c_mill.html. (1)

There are three common ways of processing dry corn, (1) a tempering degerming process, (2) stone-ground or nondegerming process, or (3) alkaline-cooked process. Each miller has his own unique variations of the overall processing system. Let's look at the most common process, that being (1) tempering degerming.

First the corn is dry cleaned to separates fines and broken corn from the whole kernels. Sometimes a wet cleaning follows to remove surface dirt, dust, and other matter. The cleaned corn is tempered to 20 percent moisture content, which eases the milling processes. While moist the outer parts of the corn are removed to leave the starchy endosperm, much the same idea as with a wheat grain only much larger in size. The endosperm comprises the bulk of the kernel and it is further processed to remove the germ. Then the endosperm is dried, cooled, and sifted. Further grinding and rolling is used to make a variety of smaller grits, meals, and flours.

This process is much more sophisticated than the process at the Dutch mill in Pella, Iowa, but they may have tempered or wet the grain before grinding it with the stones. White corn is often stone ground to produce hominy grits and corn meal, and little of the hull and germ have been removed. So you get it all honey, plus a little stone grit to polish your teeth! Gosh, the stones wear down so the grit must go somewhere, as in the meal. I am just guessing on that one.

About 65% of the corn processed emerges as prime products and the remaining 35% emerges as byproducts. Refer to the article at the web site for additional food products that are being made by corn mills. Corn is indeed a versatile feed-stock.

APPENDIX J

Oat Milling

If you were to look at a grain of oats you might wonder if it was some sort of weed seed and wouldn't suspect that beneath that lifeless hull is some oatmeal cookies just waiting to be devoured by hungry grandkids. A web site article produced by the North American Millers Association gives a good rundown on how oats are milled, so what follows is brief excerpts from that work.

Raw oats undergo cleaning to remove foreign materials such as other grains, little rocks, and who knows what else the sieves may separate out. Hulling is done to remove the hull from the meaty center, called the groat in the trade, and this process caught my eye. Somehow the oat grains are flung against a rubber ring, which knocks the hull off without damaging the tender inner meat or groat. This is a major departure from milling wheat or corn because these grains do not have this outer hull like oats (and barley) does.

The separated hulls along with the groats are sent to an aspirator step where air currents lift the light weight hulls from the stream. The remaining groats are then scoured somehow with brushes to further clean them. Then moisture is added before heating the groats to approximately 215 degrees Fahrenheit, which gives them a toasted flavor, and some other chemical processes also occur to make them safe for storage without spoiling.

A cutting step makes the groats ready for the flaking process, and more moisture is again added to increase the elasticity. Various types of oat flakes are then produced by the flaking system. A flour and bran system produces whole oat flour or a combination of bran and flour.

Web site www.namamillers.org/prd_o_mill.html (1) gives an excellent account of the oat milling process and the many products that can be made from oats. Once again, my thanks to the North American Millers' Association for their informative and easy to read web site.

Appendix K

Flour Mills In Kansas

Web site www.kswheat.com/general.asp?id=257 (1) lists a number of flour mills in the state of Kansas. The mill I am particularly interested in is a small one at Hudson, Kansas called Stafford County Flour Mills Co. which is a family owned mill that is still operating after 100 years and their prices compare very well with the two major brands I compared them to in the local grocery.

Type or Flour (5 lb bag)	Hudson Cream Prices	Brand Name Prices
Enriched unbleached	$1.33	$1.77-$2.15
100% whole wheat	$1.63	$2.77
Bread flour	$1.53	$2.15

Brand name loyalty is a great influence in choosing a brand of flour, and mom is likely to use what grandma used, but aren't the price differences enough to make you want to try the lower priced flour? The web site address for the Stafford County Mills Co. is http://www.hudsoncream.com (2), so go there and have a look at the many products this little mill produces. I can't help but wonder why their advertised prices on the web site are more than what I found in the local grocery store.

The mills in Salina and Newton are part of a major player in the grain and oil-seed business, and some of the rest I do not recognize by name. It appears that some of those listed are specialty mills and probably are relatively small in volume.

I did find a mill that processed Durum wheat, although it is located in St. Louis, Missouri rather than in Kansas. The name of the mill is US Durum Milling Inc (3). Further study reveals that this operation is owned by Archer Daniels Milling Co., which is a biggee in the milling industry

APPENDIX L

Milk Processing In Kansas (1)

Milk processing in Kansas (2003) has become centralized into two large plants that are located in Wichita and Hutchinson, which results in raw milk being shipped long distances by tank truck from the farm to the plants. It makes sense to the processor to be located near a large customer base because once the effort and expense of processing the milk has been invested the processors surely want to sell the finished products before spoilage eats into their profits. What we have then, is a lot of diesel fuel being expended to ship the raw milk to the urban plants and then to ship much of the processed milk back to the customers in the outlying regions. If Fort Hickok can have a milk processing plant that can serve the needs of thousands of people in the nearby areas then a great deal of this trucking expense can be avoided and some new jobs for rural America can be created.

To make matters worse, a dairy plant in the Kansas City area, on the far northeast corner of the state, and one in Iowa also serve major segments of the Kansas market. There are a few smaller plants in Kansas, probably making cheese, but even these small plants have to ship their cheese hundreds of miles to market after seeing the raw milk come a long way to their doors for processing. We can do it differently at the green community by producing the milk locally, processing it locally, and then limiting the sale of the output to nearby markets in order to conserve motor fuels needed to haul it to market.

If you think that *milk processing* is a mouthful then try *Value-Added Dairy Processing Feasibility Report,* which is the title of the work (1) from which most of this information has been taken. The article is a PDF, and is long, but it does a great job of covering the many aspects of working with milk and the resulting products. There are plenty of regulations for milk processing, even market orders to regulate the price of milk, but aside from that it is an interesting and labor intensive industry. The article gives a number of diagrams and discussions on how various products are made from milk.

1. WHOLE MILK CONVERSION

How much milk does it take to make butter, cheese, or ice cream? This section of the article gets right down to it, 21.2 pounds of whole milk to make a pound of butter, 10 pounds to make whole milk cheese, with cheddar cheese being one of the whole milk cheeses, 12 pounds of milk to make a gallon of ice cream. I have extracted the information and have placed it into Table 5, which is in Appendix E for your convenience. Let's consider a few products.

Most of the fluid milk becomes whole milk and 2% butterfat milk. I think of whole milk being about 3.6% butterfat or greater, and you can bet the processing plants will separate off the excess cream so it can be marketed in other forms. The input raw milk is blended, pasteurized, homogenized, and then bottled before storage in a cooler until it can be shipped out for consumption. Pasteurization kills the pathogens that may be in the raw milk, and homogenization causes the cream in the milk to not come to the top, as it was in the raw farm milk I used to drink as a kid. Wow, if mom were to skim off some of that good stuff and then pour the skim milk, or blue milk, back into the cans that were to be taken to the local cheese factory, then my dad would have a conniption because the blue milk diluted the butterfat percentage whole milk in the cans. Pop's milk check depended heavily upon the percent butterfat, and he milked small breeds that gave very rich milk relative to that given by a large breed of cow, so he was proud of being able to sell that rich creamy milk. It is a strange thing, some of the cream in the milk used to make cheddar cheese was still in the whey and was separated out and sent to another plant for use, so I cannot understand why the butterfat content was such an issue except for assuring that farmers did not dilute their milk with water before selling it to the plant. Water doesn't make much cheese they say. I used to work in a small cheese plant as a kid, so maybe I can share some of that experience with you later in this work.

2. FLUID MILK PRODUCTS

This section of the article defines whole milk as that not having less than 3.25% butterfat, and Vitamin D and A can be added to the milk for enrichment. The small cheese factory where I worked did not enrich their milk but it did sell some milk to customers who brought in their own containers. Ours was a little galvanized milk can that held maybe six quarts. By asking the customers to furnish their own container the plant avoided the expense of a bottling operation, which might be a possibility for a small plant at Fort Hickok if health regulations permit such.

Gee, 3.25% butterfat, my dad's Jerseys and cross breeds gave milk with 3.8% butterfat, and all these years I thought that was whole milk. Oh well, the industry has to favor the big producers, the Holstein herds, as they don't give as much butterfat per gallon as a small breed does. Some dairy farmers keep a few Jerseys around just to add to the butterfat content of their milk, so there must be a place for even a little Jersey out there with the big Holsteins. I understand a Holstein cow, as big as she may be, is a rather gentle animal. Pop's Jerseys were also gentle, as long as they got what they wanted!

We spend a lot of money on disposable jugs that often end up in landfills, and fuel is needed to haul these things around with the milk they hold, so let's get creative and do it another way, hopefully in a green way. While doing Kitchen Police (KP) in the service I had occasion to replace the large containers in the milk dispensers at the chow hall. These were about 5 gallon capacity and were throwaways, being made of heavy cardboard with a plastic liner and spout. Once again, more trash for the city landfill, so let's keep that in mind as we go on. I think the move to throwaway containers for milk may have been for the same reason that sodas and beer went to throwaway containers. Cleaning milk jugs or pop and beer bottles for reuse takes equipment and energy to operate, so some wise guy reasoned that a throwaway container might be more profitable. Could be, and one good thing to come of it is that the empties didn't have to be trucked back to the bottling plant, which saves some fuel. Do you think that saving fuel was a prime consideration? I doubt it. Perhaps the throwaway containers are more sanitary than the reusable ones, so let's give the corporate bozos a bit of credit anyway.

A number of delicious products can be made from the fluid milk, which is the feedstock to the processing operation. I love buttermilk, and it is like getting used to drinking tomato beer or *big red ones.* Once you get used to looking at the glass after drinking the stuff you have it made! Half and Half is great for coffee and to be used over cereal if your doctor doesn't know how you are abusing your cholestoral level. Refer to page 13 of the article (1) for a more complete listing of the many products that can be made from fluid milk, and I think that a plant in Fort Hickok could produce a number of the more popular ones such as bottled or bulk milk, creams, and cottage cheese. I will speak of making cheddar cheese later but let me say that cottage cheese, which is a relatively simple cheese to make, is made from skim milk and contains less than .5% milk fat. Oh really, that is for *dry curd cottage cheese,* and I being a glutton for punishment, bring home the good stuff, the 4% cottage cheese, because it tastes richer than the healthier versions. Cream can be measured back into the cheese to create the desired butterfat content, so tuck that little tidbit into your trivia box.

3. FLUID MILK PRODUCTION AND CONSUMPTION

This section of the article details the different uses for fluid milk in the United States. And approximately 33% of the milk from the farm moves into fluid milk products. Consumption of fluid milk products has fallen, due in part to the fact that other beverages such as beer and soda pop are readily available in the stores that sell milk. However, the total volume of milk from the farm has increased, so there is a demand somewhere for Bessie's nectar. Whole milk is the largest share of milk sales, followed by 2%, 1%, skim milk, and buttermilk, and half and half. Geez, buttermilk and half and half being grouped together. No way, don't mess with my buttermilk by doctoring it with half and half. Never fear, that is not what the statistics are saying. Sorry 'bout that little tangent.

Milk production tends to be seasonal, as the cows don't produce as much during the hot months as in the cooler ones. My father used to dry his cows up during the heat of the summer, primarily so he could concentrate on his field work, but I can also remember that the old cows didn't give much in the hotter months. I used to help milk those old girls by hand, so I should know about that. A warm cow is nice to lean into while milking in cold weather, even in a cold rain if your are milking outside in the pasture months, but the water drips off the cow and into the milk pail, diluting the butterfat. Oh my, don't tell my daddy that I was diluting the milk with runoff from my little Jersey. Wow, her milk was so rich it had a golden color from the heavy cream, well over 4% butterfat.

4. CHEESE

After making cheddar cheese as a kid I still have a special interest in making cheese. We probably did it the hard way back then, but the little plant in Tescott, Kansas made *Tescott Mellowmade* cheddar cheese, and some of it was shipped all across the country. We sometimes made large hoops, about 90 pounders, to sell to the government, and it was the color of Swiss cheese because Uncle didn't want the food coloring to be added to the cheese to give it the characteristic golden color. My memory is a bit fuzzy on this one after being away from it since the mid '50s, but it seemed that about a pint to a quart of food coloring was added to the vat of milk, which must have been on the order of 10,000 pounds (1,200 gallons). It was stirred into the vat of milk by rotating arms fitted with paddles, and the monstrous thing moved back and forth the length of the vat to give the milk a good stirring.

Table 5 of Appendix E tells us it takes 10 pounds of whole milk to make one pound of whole milk cheese, such as cheddar cheese. That sounds about right as we canned about 80 hoops of cheese per vatful, a *hoop* being a press can that yielded a 12.5 pound *longhorn* of cheddar cheese. A longhorn of cheese has a cylindrical shape that is slightly narrower at one end than the other, and the ones we made were maybe 14 inches high by six inches in diameter at the bottom or largest end. This is done so the cheese can be removed from the press can the next morning and it was a rowdy job (for a kid!) that I loved to do each morning. The press can, the *hoop,* had two handles or ears on it that enabled it to be handled while filling it with new curd in the vat and working with it on the cheese press. The cheese was removed from this can by banging the open end down on a heavy wooden block, and the longhorn, still rather wobbly at this stage, came out and was placed on a drying rack that had a frame to keep the tender cheeses upright until they firmed up in the cooler.

On occasions we also made the bricks of cheese, usually for the holiday season, and these little press cans (2.5 and 5 pound sizes) were simple to fill from the vat. Someone simply forked the curd into the waiting cans, laid side by side on the vat floor, and then a lid was placed on the can to hold the curd inside the can while it was being pressed for the night.

The pressed cheeses were sent to a cooler, which was kept in the low 40s for temperature, to age and firm up for a day before the cooler man (old Harry Pierce) trimmed them, marked them with an edible dye or ink, and then on the next day dipped them in hot paraffin so they would keep. The waxy paraffin kept the cheese from drying out or from molding. Four longhorns were placed in a cardboard box, and the date of manufacture and the gross weight of the box, about 52 pounds, was marked on the side of the carton. And the price to a milk customer, about forty six cents a pound for some darned good mellow made cheddar cheese back in the fifties, the golden age of hope for young Americans, even farm kids that worked part time in a cheese factory for 60 cents per hour. My father worked alongside me as a moonlighter for 70 cents per hour. Shoot I had plenty of spending money, and dad's gas barrel and tools to keep my old Chevy running, so what more could I want as a teenager back then. And beer was a quarter a bottle, if an older person could be talked into buying it for us whipper snappers that were under age.

The factory couldn't keep the cheese long enough to age it, as the demand was heavy for years, but the owner managed to salt a way a few boxes for aging, which made nice gifts. Wow, a 12.5 hunk of cheddar, talk about a hog's heaven. My aunts and uncles were city folks and they usually came to Tescott for Memorial Day, and good old dad treated them with a hunk of Tescott Mellowmade to take home with them.

So what happened to this little cheese plant that had pleased folks with its offerings since 1934 and meant some jobs for about eight locals? It finally closed because the corporation that bought it out after the owner retired didn't think it was profitable to continue. A sister plant at Alma, Kansas still operates, probably because it is closer to the major population centers of Topeka, Lawrence, and the Kansas City area. Energy costs may have been a major factor in the closing, because the raw milk was being trucked in for miles around and then the cheese had to be shipped at least 100 miles to a major customer in Wichita, Kansas.

Where in heck did the name of *Tescott* come from? Well it wasn't taken off a Cracker Jack box, but was related to a pioneer founder of the little town. His name was Thomas E. Scott, hence the name of Tescott.

Some of the Mellowmade cheddar cheese was processed by another firm to become American cheese, the kind you eat on your cheeseburgers. It is my understanding that the cheddar is heated to melt it and then additional cream is added to give it the creamy texture and flavor that is so popular with Americans. The article speaks of other cheeses, particularly the Italian types such as that used for pizza and pasta dishes. These are also popular types of cheese and little Fort Hickok can be a player in this market as well as in the more traditional fluid milk and cheese markets. Gourmet cheeses are a favorite for travelers, and my family and I are pushovers when it comes to buying a small hunk of something to munch on as we travel. One place in Missouri, at Osceola, even had an inexpensive cheese knife to offer, as we didn't have a knife in the car. We do now, and we just leave it in the car in case we have a cheese attack while tooling along the interstate highways.

I'll not detain you by going into the details of making cheddar cheese, but the article has a good section on the process. Just one more piece of trivia before moving on to the final topics of this appendix. *Cheddar* refers to a step in the cheese making process, and there is also an alternative to the old fashioned cheddar step that supposedly gives the same results in less time. The cheddar process refers to the practice of letting the cooked curds, which are little cubes about ¼ inch across, settle to the bottom of the vat while the whey drains away, to be ran through a cream separator to take more of the cream out of the whey. The settled curd becomes a long slab along each side of the vat gutter, which is simply the low point in the center floor of the vat where the juices (the whey) run off to a receptacle and a pump. Once the acidity of the runoff whey reaches a certain point the long slabs are cut into narrower ones, each being about 12 inches wide and 24 to 30 inches long by a few inches thick. These slabs are like big pieces of foam rubber and are stacked and restacked upon each other until the acidity is at a certain point before they are milled into large chunks, which we also called curds. These chunks, after being salted and rinsed of the excess salt, are then placed into the press cans to make the raw cheese or longhorns.

I loved to throw around the big slippery slabs of cheese curd as it continued to work or reach the desired state for salting and canning, and this old process was the cheddaring step. Now a *stirred curd* process is used to get the same result of the old cheddaring step, so why in heck are they allowed to call it cheddar cheese? This new process makes the process shorter and simpler because the little cooked curds are simply placed in the cheese press cans, or hoops, and apparently the whole process now takes 4.5 hours rather than 5.5 hours to complete the old cheddaring way. What a bummer, this new fangled process, as I would rather flip slabs of cheese curd than to play ball. Do you suppose that is why the folks at the coffee shop called me a cheese ball?

I invite you to browse through this long and informative article and you might find yourself wanting to flip some cheese someday. Oh my, that sounds tacky? Let's take a quick look at a *market order* before moving on to a topic that may be applicable to our new green city.

5. MARKET ORDER

There are so-called market orders for a variety of agricultural products, even peanuts if I remember correctly. I see it as a government mandated free-trade restriction from the standpoint of a consumer, but there is a legitimate need served by these market orders. The Milk Market Order is designed to protect the producer, the farmer, from being ripped off by the processor. In other words, one processor cannot undercut the prices paid by other processors in the area. There are different classes of milk and the formulae for determining the fixed price of milk are rather complicated. As well intended as the market order may be, I see it as a sort of protectionism that does not favor the consumer, but there is an alternative way to produce and market milk products, one which I hope would be tried at Fort Hickok. Which brings me to the next and final article, *Value-Added Dairy Options*

6. VALUE-ADDED DAIRY OPTIONS (2)

This article leans towards organic milk products, which is fine, but there is something to be learned from it, even for those who may not want to go into the organic business. I see merit in the concept, but it may tend to limit the customer base for a small plant such as the one contemplated for Fort Hickok. Dairy farmers can add value to their raw milk by doing the processing themselves, and this is actually being done in northeastern Kansas by a group. Typically, the producers form a cooperative that includes the dairy farms (Bessie and her milk), the tank trucks, and the processing plant.

The regulations are mind boggling but the effort can be profitable, given time and perseverance. And the long haul can be purposely designed out of the operation, plus the quality of locally produced products can be more closely monitored than when buying from a distant source.

Appendix M

The Mother Of All My Forts

In October of 2002 I designed a planned retail development with a historical theme that was to replace most of what was Old Abilene Town, a recreational theme park that had gone downhill due to lack of interest in the wild west theme.

The total land area occupied by this proposed little fort was 700 feet by 250 feet, or two city blocks in Abilene, Kansas, and it was to replace about two thirds of the existing theme park. Parking was to be along the perimeter and in adjacent lots because space within the boundaries of the development was at a premium and the challenge was to pack it with many attractive businesses that would appeal to tourists as well as local patrons. Not permitting vehicles on the premises allowed more of the acreage to be used for the interesting buildings that were to have a frontier military fort theme, historically speaking, but which actually offered some modern day shopping opportunities.

There was to be 20 buildings and two old timey watch towers to attract attention to the site and to provide a hint of it being a frontier fort. Two of the 20 structures were from the WW II era, and were not likely to be seen as such on an old frontier post, but I wanted to include these in order to offer the services they represented.

A parade ground soon gave way for space needed by the retail shops, but a touch of parading was preserved, including the post flag pole. This was to be a commons area for gathering on special occasions such as Independence Day or Memorial Day.

Following is a list of the structures with their military name or function and the corresponding civilian name or function.

1. Post Exchange—factory outlet stores

2. Armory—Guardhouse—western clothing, hunting and fishing items, and a gun shop with ammo sales

3. Post Headquarters—a multi purpose structure that included a museum, administrative offices for the development, public restrooms, and rental space for professional offices and retail sales. It was one of the few two story buildings on the post, which reflected the building practices on frontier forts, as lumber was scarce and the work was often done by the troopers with very rudimentary hand tools.

4. Family Housing—two duplexes, usually for the officers and their families back then.

5. Stables—Commissary—a small supermarket. Possibly a tall barn-like structure designed to provide natural ventilation for the horses. Fort Riley, Kansas once used such a structure to house the post commissary after it had been used to exercise and show horses back in early cavalry days.

6. Bachelor Officer Quarters (BOQ)—eight rental rooms with a laundry and linen storage on the back or street side. If it were used as a true BOQ then there would have been four suites for single officers that were larger than a modern motel room. These officers typically took their meals in a special mess hall so there usually was no kitchen provided in these quarters. A married but unaccompanied officer might also stay in the BOQ.

7. Post Commander's Quarters—to be used as a steak house, with a little enlargement of the kitchen area.

8. Laundress's Quarters—four rental suites with a room on the back for doing the linens needed by the motels and the steak house. Laundry maids were typically wives of noncommissioned officers and did the washing for the lower grade soldiers who lived in the barracks. Quite often she had to do the job right in her own quarters.

9. Bakery/Clothing Sales Store—In old forts these two functions were in separate buildings because of the danger of fire breaking out. The designed use in the new fort would be dual purpose, a boutique for the ladies and a bake shop.

10. Post Barber Shop—a shoe store plus a unisex barber shop

11. Sutler's Store—a post exchange or convenience store back then, but now including a bar or tavern, a fast foods restaurant, and a convenience store.

12. Mess Hall—a cafeteria

13. Post Library—probably only a room in and old fort, but in the new one it was to serve as a museum, small gift shop, and bookstore.

14. Maintenance Shed—a quartermaster function back then—still used for maintenance of the development and as a security office.

15. NCO Club—Noncommissioned Officer's Club—of the WW II era in size—a large hall for dancing and for stage entertainment.

16. Service Club—usually for the lower graders, and no alcohol being served, typical of modern military installations and sometimes referred to as the USO club. Used as a youth center in the new fort.

17. Quartermaster Stores and Tailor Shop—now selling domestic linens, bed and bath items, plus sewing and quilting supplies.

18. Post Chapel—still a chapel seating about 30 people

19. Post Hospital or Dispensary—gifts and hobby supplies

20. Canteen—coffee shop, ice cream sales and soda fountain.

APPENDIX N

Strawberry Production In The U.S. (1)

Strawberries were known to ancient Roman poets in the first century but the California industry was launched in the early 20th century. Most of today's California strawberries are hybrids of a large berry native to the cooler regions of the Pacific slope from Alaska to the tip of Chile, so the sweet thangs are world renown.

Strawberries are indeed popular in the U.S., ranking fourth behind grapes, oranges, and apples in terms of value produced. Most of our strawberries are grown in southern and coastal areas of California, and they like warm days with low humidity. Florida and Oregon follow California in strawberry production. Geez, if it takes warm days and low humidity, then welcome to Kansas Mr. Strawberry Farmer? Really now, in spite of the intense heat and drought, my mom's little patch in the deserted chicken yard just kept on a comin' until the old plants gave their all each year, so maybe the heat and drought wouldn't be a big factor except to limit the heavy producing season. Maybe that is why they invented June bearing strawberries, for the folks in Kansas, as June isn't so bad in terms of moisture and heat. Just stay away, or I should stay in, during July and August.

April and May are the peak shipping months for winter strawberries coming from Florida, which would work fine for producers in the Plains area as their berries won't be ready until later anyway. I think the key thing for a Plains strawberry farmer is to keep the distance to market as short as possible for fresh berries and to process some so there can be local berries available in the off season. Frozen berries and products like jams, jellies, and glaze would help to prolong the demand for local berries into out of season months. And strawberry shortcake, I will take it any way I can get it, frozen, fresh, or whatever, just bring it on! I see this as being a special treat in the areas where berries are being produced, something that one doesn't get very often in a restaurant.

The article referenced by Note 1 of Appendix N gives a comprehensive look at strawberry production in the U.S., including demand per capita, which is over four pounds per person for fresh berries and about one pound per person for frozen ones. Have you had your strawberries today?

If most of the strawberries are produced in Florida or in Ventura and Monterey counties of southern California, then what are we looking at in terms of miles to the Kansas City, Kansas distribution center that serves the Abilene grocery? The road miles from Salinas, California is 1901 miles and the driving distance from central Florida (Orlando) would be 1241 miles. Given this distance and a truck using diesel fuel at the rate of 5 miles per gallon, a load of strawberries would take 380 gallons to make the trip from California and 248 gallons from Orlando, Florida. Compare this to a trip of 100 miles or less from a local berry farm with a truck that gets better mileage, using something other than diesel fuel, and I think you can see where growing strawberries locally might be in the best interests of our national energy posture. Hopefully it would also be easier on your pocketbook. More berries for da buck, that's my motto. Yes, some of the local demand will still be satisfied with trucked in berries, but why not serve what we can from our own farms? And besides, the grazers will appreciate a few left on the plants, as will the wildlife, but not if I get there first, those belly robbers.

Notes

Chapter 1 Traditional and New Field Crops

1. Dove's Farm Foods Ltd, Berkshire, U.K. (n.d.). *Organic Grain, Wheat and Flour—About Flour—The Composition of a Grain of Wheat.* Retrieved June 1, 2007 from
http://www.dovesfarm-organic.co.uk/composition-grain-of-wheat.htm

2. Wikipedia Foundation, Inc. (n.d.). *Barley.* Retrieved May 30, 2007 from
http://www.wikipedia.org/wiki/Barley

3. Western Area Power Administration, (Vol. 23, No. 2, April 2004). *Faith in mustard seed fuels Blue Sun's vision of biodiesel cartel.* Retrieved June 9, 2007 from http://www.wapa.gov/es/pubs/esb/2004/april/apr048.htm

4. Green, Donald, Forage Specialist. Manitoba Agriculture and Food, Soils and Crops Branch. *Switchgrass as a Biofuel—Is it Economically Feasible?.* Retrieved April 30, 2003 from
www.gov.mb.ca/agriculture/news/topics/daa26d01.html. This reference is no longer available (June 2007) but the percentage ash remaining after combustion is in line with other sources.

5. Bransby, David. (n.d.) *Switchgrass Profile.* Retrieved June 9, 2007 from Auburn University Web site: http://bioenergy.ornl.gov/papers/misc/switchgrass-profile.html.

6. Questions & Answers about Miscanthus. (n.d.). Retrieved June 9, 2007 from http://bioenergy.ornl.gov/papers/miscanthus/miscanthus.html

Chapter 2 Truck Garden Crops

1. Mapquest.com is a proprietary internet product that helps people to find directions to a location and gives the mileage from one point to another. MapQuest, Inc. is a wholly owned subsidiary of America Online, Inc. To use

the program key in www.mapquest.com and follow the instructions on the resulting web page.

2. Walton Feed, Inc, Montpelier, ID. *(April 10, 2000). Water Content of Foods.* Retrieved June 9, 2007 from www.waltonfeed.com/self/h2ocont.html

3. USDA—National Agricultural Statistics Service (NASS). PDF. Crop Years 2002 thru 2004. The yield of food grains can vary from year to year so I have chosen an average or typical value for some of the products.

4. Boriss, Hayley, Brunke, Henrich, & Kreith, Marcia. (March 2006). Agricultural Marketing Resource Center, University of California. *Commodity Profile: Strawberries.* Retrieved June 10, 2007 from http://aic.ucdavis.edu/profiles/Strawberries-2006.pdf

5. Mushroom Industry Report (94003), tab10.xls. (May 2003). *Table 10—Mushrooms: Number of growers, yield;, and dollar volume, 1966/67-2001,0 21/.* Retrieved from Web site: http://usda.mannlib.cornell.edu/data-sets/specialty/94003/tab10.xls

    The above web site is by the Albert R. Mann Library, of Cornell University, Ithaca, NY and the article was apparently part of those that a man named Richard Clarke used for a research project.

Chapter 3 Greenhouse Products

1. Marr, Charles W. *Commerical Greenhouse Production: Greenhouse Tomatoes.* Kansas State University. (February 1995). Retrieved June 12, 2007 from http://www.oznet.ksu.edu/library/hort2/MF2074.PDF

2. USDA, Agricultural Research Service. (Updated June 24, 2005). *Greenhouse-Grown Bell Pepper Production.* Retrieved June 12, 2007 from http://www.ars.usda.gov/is/np/mba/jun05/pepper.htm

Chapter 4 Agribusiness Industries

1. Jackson Frozen Food Locker, 400 South High,Jackson, Missouri 63755. (n.d.). *Beef Carcass Breakdown.* Retrieved June 12, 2007 from www.askthemeatman.com/yield_on_beef_carcass.htm

2. Jackson Frozen Food Locker, 400 South High, Jackson, Missouri 63755.(n.d.). *Interactive Pork Chart*. Retrieved June 12, 2007 from www.askthemeatman.com/hog_cuts_interactive_chart.htm

3. Chapman, Frank A., Assistant Professor. *Farm-raised Channel Catfish.* (Circular 1052, first Published July 1992). Department of Fisheries and Aquatic Sciences, Cooperative Extension Service, Institute of Food and Agricultural Services, University of Florida, Gainesville, 32611. Retrieved June 12, 2007 from http://edis.ifas.ufl.edu/FA010

Chapter 5 People Oriented Businesses

1. Rutlader Outpost Web site: http://www.rutladeroutpost.com/. Retrieved June 10, 2007. Click on the Middle Creek Theatre button to see more about the opry.

2. Pella, Iowa Web site: http://www.pella.org/. Retrieved June 10, 2007.

3. Amana Colonies Web site: http://www.amanacolonies.org/. Retrieved June 10, 2007.

4. Sandy Lake ParkWeb site: http://www.sandylake.com/. Retrieved June 10, 2007.

5. Wakeman, Scott. *Auto Detailing: Tips an Tricks for the "Driveway Detailer"!. (n.d.).* Web site: http://www.corral.net/tech/maintenance/detail.html Retrieved June 10, 2007.

Chapter 10 Communal Living Concepts

1. Meunier, Rachel. *The Farm. (12-17-94).* Retrieved June 11, 2007 from www.thefarm.org/lifestyle/cmnl.html.

2. Wikipedia Foundation, Inc. *Residential community.* (last modified March 23, 2007). Retrieved June 11, 2007 from http://www.wikipedia.org/wiki/Residential_community

3. Wikipedia Foundation, Inc. *Ecovillage.* (last modified June 10, 2007). Retrieved June 11, 2007 from http://www.wikipedia.org/wiki/Ecovillage

4.   Wikipedia Foundation, Inc. *Intentional community.* (last modified May 21, 2007). Retrieved June 11, 2007 from http://www.wikipedia.org/wiki/Intentional_community

5.   Wikipedia Foundation, Inc. *Cohousing.* (last modified May 27, 2007). Retrieved June 11, 2007 from http://www.wikipedia.org/wiki/Cohousing

6.   Wikipedia Foundation, Inc. *Cooperative.* (last modified June 12, 2007). Retrieved June 11, 2007 from http://www.wikipedia.org/wiki/Cooperative

Appendix A Fort Hickok, Whre Are You?

1.   Caterpillar web site: www.cat.com/cda/layout Retrieved June 10, 2007.

Once you reach the home page of Caterpillar then click on Products, then Power Generation, then Generator Sets. The page reached by clicking on Power Generation shows some yellow engines, V 16's. I once had an occasion to see one of these monsters at Larned, Kansas, which is near Fort Larned, a historical frontier fort, and the generator unit was running. I touched the generator while the system was powered up and what vibes it had!

2.   Ingersoll-Rand web site: http://www.ingersollrand.com/ Retrieved June 10, 2007.

Appendix B Production? How Much is enough?

1.   United States Census Bureau. *State and County Quick Facts (Kansas).* (Revised May 7, 2007). Retrieved June 10, 2007 from http://quickfacts.census.gov/qfd/states/20000.html.

I selected the counties, one at a time, and added up the population of these counties. By clicking on the Kansas County Selection Map button you can see a map of Kansas with the counties shown. The quadrants I have defined for my work are separated right from left by a line running along the east side of Smith County, on the north side of the state, to a line running along the east side of Barber County on the south side of the state, and by a west to east line that crosses Greeley County at the western border runs to cross Miami County at the eastern border of the state.

Appendix C Housing That is Right For Your Needs and Tastes

1.   The Brown Home was built by C.L. Brown of Abilene, Kansas, and since his death a foundation continues to manage the home. See web site http:// www.emporia.edu/business/kbhfhistdetail.php?k_id=17 for a discussion of this talented and generous man. Retrieved June 10, 2007.

Appendix D Methods For Making Biofuels

1.   Wikipedia online encyclopedia. *Cellulosic ethanol*. (Last Modified June 17, 2007). Retrieved June 10, 2007 from
     http://en.wikipedia.org/wiki/Cellulosic_ethanol

2.   Clean Fuels Development Coalition. *Methanol. (n.d.)*. Retrieved June 10, 2007 from http://www.ethanol-gec.org/clean/cf05.htm

3.   U.S. Department of Energy. *Gasification Technology R&D*. (Updated March 14, 2007). Retrieved from www.fe.doe. gov/programs/powersystems/gasification/index.html

4.   Wikipedia online encyclopedia. *Butanol fuel*. (Last Modified June 15, 2007). Retrieved June 10, 2007 from http://en.wikipedia.org/wiki/Biobutanol

5.   Fernando, Berton. *Hydrolysis*. ©1995, 2007. Conversion Technologies. California Integrated Waste Management Board. Retrieved June 10, 2007 from www.ciwmb.ca.gov/Organics/Conversion/Hydrolysis

6.   U.S. Department of Energy. *Pyrolysis and Other Thermal Processes*. (Last Updated October 13, 2005). Retrieved from www.eere.energy.gov/biomass/ pyrolysis.html

7.   California Energy Commission. *Anaerobic Digestion. (Last Updated April 27, 2005)*. Retrieved June 10, 2007 from http://www.energy.ca. gov/development/biomass/anaerobic.html

8.   Wikipedia online encyclopedia. *Partial oxidation*. (Last Updated May 4, 2007). Retrieved June 10, 2007 from http://en.wikipedia.org/wiki/Partial_oxidation

9.   Market Wire (May, 2006). *Syngas International Corp (OTC BB: SYNI) Ethanol Research to Focus on Cellulose Based Feed Stocks.* Retrieved June 10, 2007 from http://findarticles.com/p/articles/mi_pwwi/is_200605/ai_n16350160

Appendix F Small Community Wastewater Treatment

1.   Doley, Todd M. & Kerns, Waldon R. Virginia Cooperative Extension. *Individual Homeowners & Small Community Wastewater Treatment & Disposal Options.* (Publication Number 448-406, June 1996). Retrieved from www.ext.vt.edu/pubs/waterquality/448-406/448-406.html

Appendix G Aermotor Windmill

1.   Aermotor Windmill Company. (Copyright 2002-2006). Retrieved from www.aermotorwindmill.com on June 10, 2007.

2.   Google is a program product of the Google Corp and is a search engine that is used to find items by entering a search argument value and then pressing the *Enter* key. See www.google.com

Appendix H Wheat Milling

1.   North American Millers' Association. *How Wheat Flour Is Milled.* (Copyright 2006). Retrieved June 11, 2007 from www.namamillers.org/ci_products_wheat_mill.html

2.   Kansas Association Of Wheat Growers. *How Wheat Becomes Flour.* (n.d.). Retrieved June 11, 2007 from www.wheatmania.com/allaboutwheat/wheatfacts/howwheatbecomesflour.htm

3.   The Stafford County Flour Mills Co. *About Us.* (n.d.). Retrieved June 11, 2007 from www.hudsoncream.com/about.taf

4.   Struempler, Barbara, Dr., Nutritionist. Alabama Cooperative Extension Unit, News and Public Affairs Unit. (2001 Archive, Auburn, Dec. 3). *Using Right Kind Of Flour Important To Holiday Baking.* Retrieved June 11, 2007 from www.aces.edu/dept/extcomm/newspaper/dec3c01.html This web site address can be balky so if all else fails use Google to search for 'Using Right Kind of Flour Important to Holiday Baking'.

Appendix I Corn Milling

1. North American Millers' Association. *Corn Milling Process.* (Copyright 2006). Retrieved June 11, 2007 from www.namamillers.org/prd_c_mill.html

Appendix J Oat Milling

1. North American Millers' Association. *The* Oat *Milling Process.* (Copyright 2006). Retrieved June 11, 2007 from www.namamillers.org/prd_o_mill.html

Appendix K Flour Mills In Kansas

1. Kansas Wheat Commission. *Flour Mills In Kansas.* (Copyright 2003). Retrieved June 11, 2007 from www.kswheat.com/general.asp?id=257

2. The Stafford County Flour Mills Co. *About Us. (n.d.).* Retrieved June 11, 2007 from www.hudsoncream.com/about.taf

3. U S Durum Milling Inc. 7900 Van Buren St, St Louis, MO 63222

Appendix L Milk Processing In Kansas

1. Hammarlund, Ray. Cooperative Development Specialist. Kansas Department of Commerce, Agricultural Marketing Division. *Value-Added Dairy Processing Feasibility Report.* "A Catalyst For Thought". (July 2003). Retrieved June 13, 2007 from www.agmrc.org/NR/rdonlyres/A6DF2F44-76A7-481D-93BC-2FD804EA499E/0/dairyprocessingreport.pdf

2. Gegner, Lance E., Agriculture Specialist. National Center For Appropriate Technology, USDA. *Value-Added Dairy Options. (August 2001).* Retrieved June 13, 2007 from http://www.attra.org/attra-pub/PDF/valueaddeddairy.pdf

Appendix N Strawberry Production In The U.S.

1. Boriss, Hayley, Brunke, Henrich, & Kreith, Marcia. Agricultural Marketing Resource Center, University of California. *Commodity Profile: Strawberries.* (March 2006). Retrieved June 10, 2007 from http://aic.ucdavis.edu/profiles/Strawberries-2006.pdf

Bibliography

Amana Colonies. (Copyright 2007). Amana Colonies Convention & Visitors Bureau. Amana, Iowa. Web site: http://www.amanacolonies.org/. Retrieved June 10, 2007.

Aermotor Windmill Company. (Copyright 2002-2006). San Angelo, Texas. Retrieved from www.aermotorwindmill.com on June 10, 2007

Boriss, Hayley, Brunke, Henrich, & Kreith, Marcia. (March 2006). *Commodity Profile: Strawberries.* (March 2006). Agricultural Marketing Resource Center, University of California. Davis, California. Retrieved June 10, 2007 from http://aic.ucdavis.edu/profiles/Strawberries-2006.pdf

Bransby, David. *Switchgrass Profile.* (n.d.). Auburn, Alabama. Retrieved June 9, 2007 from Auburn University Web site: http://bioenergy.ornl.gov/papers/misc/switchgrass-profile.html

The Brown Home was built by C.L. Brown of Abilene, Kansas, and since his death a foundation continues to manage the home. See web site http://www.emporia.edu/business/kbhfhistdetail.php?k_id=17 for a discussion of this talented and generous man. Retrieved June 10, 2007.

California Energy Commission. *Anaerobic Digestion. (Last Updated April 27, 2005).* Sacramento, California. Retrieved June 10, 2007 from http://www.energy.ca.gov/development/biomass/anaerobic.html

Caterpillar web site: www.cat.com/cda/layout. (Copyright 2007). Peoria, Illinois. Retrieved June 10, 2007. Once you reach the home page of Caterpillar then click on Products, then Power Generation, then Generator Sets.

Chapman, Frank A., Assistant Professor. *Farm-raised Channel Catfish.* (Circular 1052, first Published July 1992). Department of Fisheries and Aquatic Sciences, Cooperative Extension Service, Institute of Food and Agricultural Services, University of Florida, Gainesville, Florida .32611. Retrieved June 12, 2007 from http://edis.ifas.ufl.edu/FA010

Clarke, Richard. Mushroom Industry Report (94003), tab10.xls. (May 2003). *Table 10—Mushrooms: Number of growers, yield;, and dollar volume, 1966/67-2001,0 21/.* Retrieved from Web site: http://usda.mannlib.cornell.edu/data-sets/specialty/94003/tab10.xls
The above web site is by the Albert R. Mann Libray, of Cornell University, Ithaca, NY.

Clean Fuels Development Coalition. *Methanol. (n.d.).* Bethesda, Maryland. Retrieved June 10, 2007 from http://www.ethanol-gec.org/clean/cf05.htm

Doley, Todd M. & Kerns, Waldon R. Virginia Cooperative Extension. *Individual Homeowners & Small Community Wastewater Treatment & Disposal Options.* (Publication Number 448-406, June 1996). Blackburg, Virginia. Retrieved from www.ext.vt.edu/pubs/waterquality/448-406/448-406.html

Dove's Farm Foods Ltd, Berkshire, U.K. *Organic Grain, Wheat and Flour—About Flour—The Composition of a Grain of Wheat.* (n.d.). Hungerford, Berkshire, U.K. Retrieved June 1, 2007 from http://www.dovesfarm-organic.co.uk/composition-grain-of-wheat.htm

Fernando, Berton. *Hydrolysis.* (©1995, 2007). Conversion Technologies. California Integrated Waste Management Board. Sacramento, California.Retrieved June 10, 2007 from www.ciwmb.ca.gov/Organics/Conversion/Hydrolysis

Gegner, Lance E., Agriculture Specialist. National Center For Appropriate Technology, USDA. *Value-Added Dairy Options. (August 2001).* Fayetteville, Arkansas. Retrieved June 13, 2007 from http://www.attra.org/attra-pub/PDF/valueaddeddairy.pdf

Google is a program product of the Google Corp and is a search engine that is used to find items by entering a search argument value and then pressing the *Enter* key. See www.google.com

Green, Donald, Forage Specialist. Manitoba Agriculture and Food, Soils and Crops Branch. *Switchgrass as a Biofuel—Is it Economically Feasible?.* Retrieved

April 30, 2003 from www.gov.mb.ca/agriculture/news/topics/daa26d01.html. This reference is no longer available (June 2007).

Hammarlund, Ray. Cooperative Development Specialist. Kansas Department of Commerce, Agricultural Marketing Division. *Value-Added Dairy Processing Feasibility Report. "A Catalyst For Thought"*. (July 2003). Manhattan, Kansas. Retrieved June 13, 2007 from www.agmrc.org/NR/rdonlyres/A6DF2F44-76A7-481D-93BC-2FD804EA499E/0/dairyprocessingreport.pdf

Ingersoll-Rand Web site. © Ingersoll-Rand Company Limited. Hamilton HM 11, Bermuda. Web site: http://www.ingersollrand.com/ Retrieved June 10, 2007.

Jackson Frozen Food Locker. *Beef Carcass Breakdown*. (n.d.). Jackson, Missouri. Retrieved June 12, 2007 from www.askthemeatman.com/yield_on_beef_carcass.htm

Jackson Frozen Food Locker. *Interactive Pork Chart*. (n.d.). Jackson, Missouri. Retrieved June 12, 2007 from www.askthemeatman.com/hog_cuts_interactive_chart.htm

Kansas Association of Wheat Growers. *How Wheat Becomes Flour*. (n.d.). Manhattan, Kansas. Retrieved June 11, 2007 from http://www.wheatmania.com/allaboutwheat/wheatfacts/howwheatbecomesflour.htm

Kansas Wheat Commission. *Flour Mills In Kansas*. (Copyright 2003). Manhattan, Kansas. Retrieved June 11, 2007 from www.kswheat.com/general.asp?id=257

Mapquest.com is a proprietary internet product that helps people to find directions to a location and gives the mileage from one point to another. MapQuest, Inc. is a wholly owned subsidiary of America Online, Inc. To use the program key in www.mapquest.com and follow the instructions on the resulting web page.

Market Wire *Syngas International Corp (OTC BB: SYNI) Ethonol Research to Focus on Cellulose Based Feed Stocks*. (May 2006). Toronto, Canada. Retrieved June 10, 2007 from http://findarticles.com/p/articles/mi_pwwi/is_200605/ai_n16350160

Marr, Charles W. *Commerical Greenhouse Production: Greenhouse Tomatoes.* Kansas State University. (February 1995). Manhattan, Kansas. Retrieved June 12, 2007 from http://www.oznet.ksu.edu/library/hort2/MF2074.PDF2.

Meunier, Rachel. *The Farm. (12-17-94).* Sumertown, Tennessee. Retrieved June 11, 2007 from www.thefarm.org/lifestyle/cmnl.html

North American Millers' Association. *Corn Milling Process.* (Copyright 2006). Washington, D.C. Retrieved June 11, 2007 from www.namamillers.org/prd_c_mill.html

North American Millers' Association. *How Wheat Flour Is Milled.* (Copyright 2006). Washington, D.C. Retrieved June 11, 2007 from www.namamillers.org/ci_products_wheat_mill.html

North American Millers' Association. *The* Oat *Milling Process.* (Copyright 2006). Washington, D.C. Retrieved June 11, 2007 from www.namamillers.org/prd_o_mill.html

Pella, Iowa. Pella Convention and Visitors Bureau.(Copyright 2006). Web site: http://www.pella.org/ Retrieved June 10, 2007.

Questions & Answers about Miscanthus. (n.d.). Oak Ridge National Laboratory. Oak Ridge, Tennessee. Retrieved June 9, 2007 from http://bioenergy.ornl.gov/papers/miscanthus/miscanthus.html

Rutlader Outpost. (Copyright 2001-2005). Louisburg, Kansas. Web site: http://www.rutladeroutpost.com/. Retrieved June 10, 2007. Click on the Middle Creek Theatre button to see more about the opry.

Sandy Lake Park, 1800 Sandy Lake Road. Carrollton, TX 75006. Web site: http://www.sandylake.com/. Retrieved June 10, 2007.

The Stafford County Flour Mills Co. *About Us. (n.d.).* Hudson, Kansas. Retrieved June 11, 2007 from www.hudsoncream.com/about.taf

Struempler, Barbara, Dr., Nutritionist. Alabama Cooperative Extension Unit, News and Public Affairs Unit. *Using Right Kind Of Flour Important To Holiday Baking.* (2001 Archive, Auburn, Dec. 3). Auburn University, Alabama. Retrieved June 11, 2007 from www.aces.edu/dept/extcomm/newspaper/dec3c01.html
This web site address can be balky so if all else fails use Google to search for *'Using Right Kind of Flour Important to Holiday Baking'.*

United States Census Bureau. *State and County Quick Facts (Kansas).* (Revised May 7, 2007). Retrieved June 10, 2007 from http://quickfacts.census.gov/qfd/states/20000.html

USDA—National Agricultural Statistics Service (NASS). *2005 Agricltural Statistics.* PDF. Crop Years 2002 thru 2004. Web site: http://www.usda.gov/nass/pubs/agstats.htm

USDA, Agricultural Research Service. (Updated June 24, 2005). *Greenhouse-Grown Bell Pepper Production.* Retrieved June 12, 2007 from http://www.ars.usda.gov/is/np/mba/jun05/pepper.htm

U.S. Department of Energy. *Gasification Technology R&D.* (Updated March 14, 2007). Retrieved from www.fe.doe.gov/programs/powersystems/gasification/index.html

U.S. Department of Energy. *Pyrolysis and Other Thermal Proce*sses. (Last Updated October 13, 2005). Retrieved from www.eere.energy.gov/biomass/pyrolysis.html

U S Durum Milling Inc. 7900 Van Buren St, St Louis, MO 63222

Wakeman, Scott. *Auto Detailing: Tips an Tricks for the "Driveway Detailer"!. (n.d.).* Danbury, Connecticut. Web site: http://www.corral.net/tech/maintenance/detail.html Retrieved June 10, 2007.

Walton Feed, Inc. *Water Content of Foods.* (April 10,2000). Montpelier, Idaho. Retrieved June 9, 2007 from www.waltonfeed.com/self/h2ocont.html

Western Area Power Administration. *Faith in mustard seed fuels Blue Sun's vision of biodiesel cartel.* (Vol. 23, No. 2, April 2004). Lakewood, Colorado. Retrieved June 9, 2007 from http://www.wapa.gov/es/pubs/esb/2004/april/apr048.htm

Wikipedia Foundation, Inc. *Barley.* (Updated June 11, 2007). St. Petersburg, Florida. Retrieved June 17, 2007 from http://www.wikipedia.org/wiki/Barley

Wikipedia online encyclopedia. *Butanol fuel.* (Last Modified June 15, 2007). St. Petersburg, Florida. Retrieved June 10, 2007 from http://en.wikipedia.org/wiki/Biobutanol

Wikipedia Foundation Inc. *Cellulosic ethanol.* (Last Modified June 17, 2007). St. Petersburg, Florida. Retrieved June 10, 2007 from http://en.wikipedia.org/wiki/Cellulosic_ethanol

Wikipedia Foundation Inc. *Partial oxidation.* (Last Updated May 4, 2007). St. Petersburg, Florida. Retrieved June 10, 2007 from http://en.wikipedia.org/wiki/Partial_oxidation

Wikipedia Foundation, Inc. *Residential community.* (last modified March 23, 2007). St. Petersburg, Florida. Retrieved June 11, 2007 from http://www.wikipedia.org/wiki/Residential_community

Wikipedia Foundation, Inc. *Ecovillage.* (last modified June 10, 2007). St. Petersburg, Florida. Retrieved June 11, 2007 from http://www.wikipedia.org/wiki/Ecovillage

Wikipedia Foundation, Inc. *Intentional community.* (last modified May 21, 2007). St. Petersburg, Florida. Retrieved June 11, 2007 from http://www.wikipedia.org/wiki/Intentional_community

Wikipedia Foundation, Inc. *Cohousing.* (last modified May 27, 2007. St. Petersburg, Florida. Retrieved June 11, 2007 from http://www.wikipedia.org/wiki/Cohousing

Wikipedia Foundation, Inc. *Cooperative.* (last modified June 12, 2007). St. Petersburg, Florida. Retrieved June 11, 2007 from http://www.wikipedia.org/wiki/Cooperative

978-0-595-45600-0
0-595-45600-6